ROCKS, MINERALS
& THE CHANGING EARTH

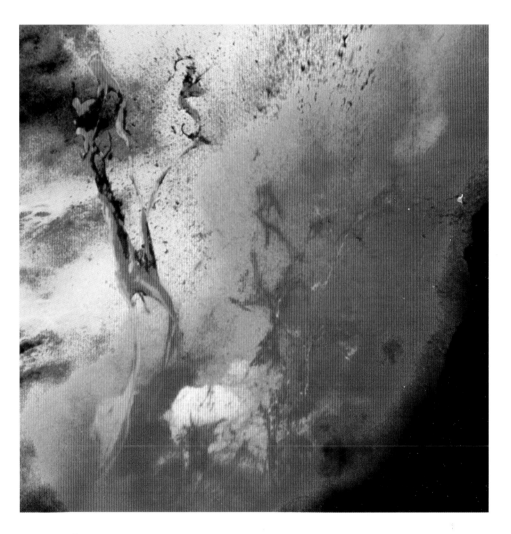

Jack Challoner, John Farndon & Rodney Walshaw

Introduction by Chris Oxlade

This edition is published by Southwater

Southwater is an imprint of Anness Publishing Ltd
Hermes House, 88-89 Blackfriars Road,
London SE1 8HA
tel. 020 7401 2077; fax 020 7633 9499
www.southwaterbooks.com; info@anness.com

© Anness Publishing Ltd 2004

UK agent: The Manning Partnership Ltd,
6 The Old Dairy, Melcombe Road,
Bath BA2 3LR; tel. 01225 478444;
fax 01225 478440; sales@manning-partnership.co.uk

UK distributor: Grantham Book Services Ltd,
Isaac Newton Way, Alma Park Industrial Estate,
Grantham, Lincs NG31 9SD; tel. 01476 541080;
fax 01476 541061; orders@gbs.tbs-ltd.co.uk

North American agent/distributor: National Book
Network, 4501 Forbes Boulevard, Suite 200, Lanham,
MD 20706; tel. 301 459 3366; fax 301 429 5746;
www.nbnbooks.com

Australian agent/distributor: Pan Macmillan Australia,
Level 18, St Martins Tower, 31 Market St, Sydney,
NSW 2000; tel. 1300 135 113; fax 1300 135 103;
customer.service@macmillan.com.au

New Zealand agent/distributor: David Bateman Ltd,
30 Tarndale Grove, Off Bush Road, Albany, Auckland;
tel. (09) 415 7664; fax (09) 415 8892

All rights reserved. No part of this publication may be
reproduced, stored in a retrieval system, or transmitted
in any way or by any means, electronic, mechanical,
photocopying, recording or otherwise, without the
prior written permission of the copyright holder.

A CIP catalogue record for this book is available
from the British Library.

Publisher: Joanna Lorenz
Editors: Joanne Hanks, Molly Perham,
Jenni Rainford and Elizabeth Woodland
Production Controller: Darren Price
Consultants: Dr Sue Bowler, Dr Bob Symes
and Rodney Walshaw
Designers: Caroline Grimshaw,
Ann Samuel and Margaret Sadler
Jacket Design: Balley Design Associates
Photographer: John Freeman, Don Last
and Paul Bricknell
Stylist: Melanie Williams and Jane Coney
Picture Researcher: Caroline Brooke
and Daniella Marceddu
Illustrator: Peter Bull Art Studio
and Guy Smith

Previously published in two
separate volumes,
Investigations: Rocks and Minerals
and *Investigations: Planet Earth*

10 9 8 7 6 5 4 3 2 1

CONTENTS

Introduction   4

ROCKS & MINERALS   6
The materials of earth   8
Looking at rocks   10
What are minerals?   12
Crystals   14
Making crystals   16
Igneous rocks   18
Making igneous rocks   20
Sedimentary rocks   22
Rocks in layers   24
Making sedimentary rocks   26
Metamorphic rocks   28
Geologists at work   30
Identifying minerals   32
Reading the rocks   34
Seeing inside the Earth   36
Limestone landscapes   38
Rock, weather and soil   40
What is soil made of?   42
Preserved in stone   44

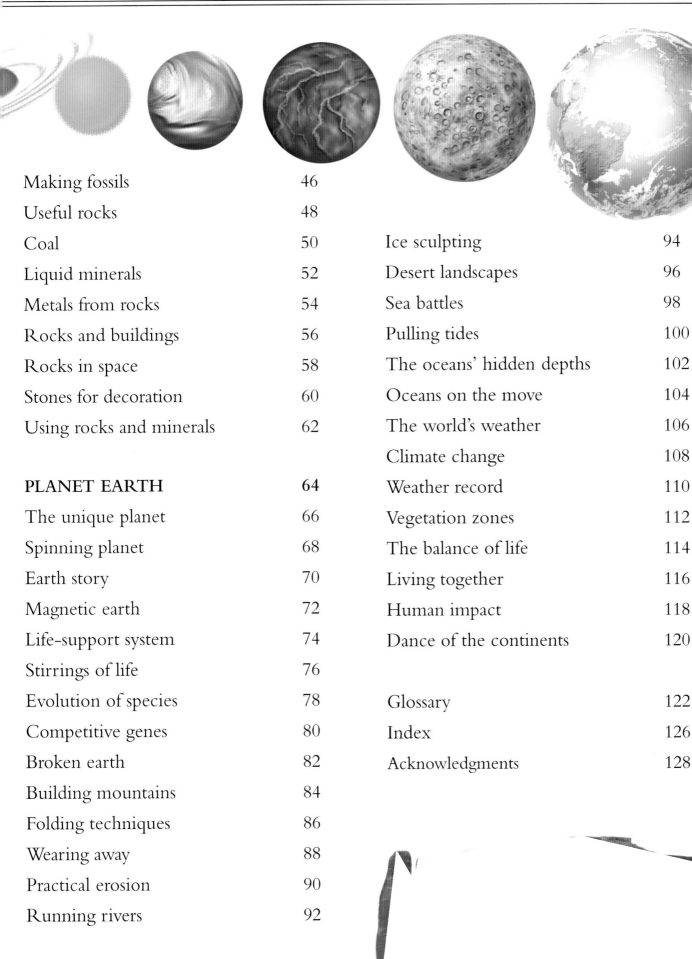

Making fossils	46
Useful rocks	48
Coal	50
Liquid minerals	52
Metals from rocks	54
Rocks and buildings	56
Rocks in space	58
Stones for decoration	60
Using rocks and minerals	62

PLANET EARTH 64

The unique planet	66
Spinning planet	68
Earth story	70
Magnetic earth	72
Life-support system	74
Stirrings of life	76
Evolution of species	78
Competitive genes	80
Broken earth	82
Building mountains	84
Folding techniques	86
Wearing away	88
Practical erosion	90
Running rivers	92
Ice sculpting	94
Desert landscapes	96
Sea battles	98
Pulling tides	100
The oceans' hidden depths	102
Oceans on the move	104
The world's weather	106
Climate change	108
Weather record	110
Vegetation zones	112
The balance of life	114
Living together	116
Human impact	118
Dance of the continents	120
Glossary	122
Index	126
Acknowledgments	128

INTRODUCTION

OUR EARTH is a unique place. Its oceans and atmosphere make it the only known planet in the Solar System where life can survive. The Earth's surface seems a solid and permanent place to live. However, it is only a thin crust of rock on a huge hot, semi-molten ball, and it has been constantly changing since the Earth was formed more than four billion years ago.

New rocks are being made all the time at volcanoes and on sea beds. At the same time, old rocks are being destroyed. They are melted deep inside the crust and worn away on the surface by the eroding effects of wind, rain, rivers, glaciers and ocean waves crashing on the shore.

Above: Ocean waves and currents erode rocky shores and beaches. These rock formations are made of different types of rock, such as granite, limestone and sandstone.
Below: Glaciers move slowly, removing all loose rocks from the surface which they pass over, leaving bare rock when the ice melts.

Rocks are a store of evidence from the past. Fossils buried in them tell us about plants and animals that lived on the Earth in the distant past. From fossils we know that millions of species of animals and plants evolved from the first simple organisms, and that extraordinary animals such as dinosaurs that once roamed the Earth have become extinct.

Our own species, the humans, have dominated the world for thousands of years. Rocks and minerals are an important resource for us. We use them for building, for roads and for decoration, and we get precious gems, oil, coal and many raw materials from them.

Our activities, such as farming and building cities, have made enormous changes to the Earth's surface. However, we are still powerless to prevent the changes caused by natural events such as volcanoes, earthquakes and floods.

Above: Sandstone rocks are sedimentary rocks made from small grains of the minerals quartz and feldspar. They form layers and are often used as building stones.

Above: Torres del Pain, Chile, was created over three million years ago. The withdrawal of glacier ice, brought the granite peaks and towers into view, which soar over 9000ft above sea level.
Right: An earthquake triggered the eruption of Mount St Helens in Washington, USA. The eruption lasted for nine hours, changing the surrounding landscape instantly, and resulting in 150sq miles of forest being destroyed.

ROCKS & MINERALS

THE MATERIALS OF EARTH

Rocks and minerals are the naturally occurring materials that make up planet Earth. We can see them all around us—in mountains, cliffs, river valleys, beaches and quarries. Rocks are used for buildings, and many minerals are prized as jewels. Most people think of rocks as hard and heavy, but soft materials, such as sand, chalk and clay, are also considered to be rocks.

Minerals make up a part of rocks in the way that separate ingredients make a cake. About 3,500 different minerals are known to exist, but only a few hundred of these are common. Most minerals are solid but a few are liquid or gas. Water, for example, is a liquid mineral, and many other minerals are found in liquid petroleum.

You might think that rocks last for ever, but they do not. Slowly, over thousands, even millions, of years they are naturally recycled, and the minerals that occur in a rock are moved from one place to another and form new rock.

Gemstones
Minerals prized for their beauty and rarity are called gemstones. The brilliant sparkle of a diamond appears when it is carefully cut and polished (as above).

pyrite

kyanite

copper

yellow sulfur crystals growing on kaolin

opal in ironstone

Minerals
All rocks are made up of one or more minerals. Minerals are natural, solid, nonliving substances. Five different minerals are shown here. Each one has definite characteristics, such as its shape and color, that distinguish it from all others. Many types of minerals are found in thousands of different types of rock.

The Earth's crust

The Earth's surface is a thin, hard, rocky shell called the crust. There are two kinds of crust—oceanic crust (under the oceans) and continental crust (the land). The recycling of the rocks that form the crust has been going on for 4,000 million years.

sandstone

granite

At some places in the hot parts of the Earth beneath the crust, huge pockets of molten rock or magma form. The magma rises, cools and solidifies to form igneous rocks such as granite. If magma reaches the surface of the Earth, it erupts as lava.

Sedimentary rocks, such as this sandstone, form from the fragments of other rocks that have been broken down by the action of rain, snow, ice and air. The fragments are carried away by wind or water and settle in a different place.

gneiss

Sometimes, within the Earth, the heat and pressure become so strong that the rocks twist and buckle and new minerals grow in them. The new rocks are called metamorphic rocks. This gneiss is a good example.

Fossils

As the fragments of rock settle in their eventual resting place, they may bury the animals and plants that lived there. The remains then become preserved as fossils. This ammonite was once a living creature, but is now made entirely of minerals.

Crystals

Minerals usually grow in regular shapes called crystals. When mineral-rich water fills a crack or cavity in a rock, veins and geodes may form. A geode is a rounded rock with a hollow center lined with crystals. The beautiful crystal lining is revealed when it is split open. Geodes are highly prized by mineral collectors.

Getting at rocks

We use rocks in many ways, but getting them out of the ground can be difficult. Explosives are often used to blast rocks out of cliff faces. Here limestone is being blasted from a quarry. Above the quarry face, large machines have been used to drill a line of holes into the ground. The holes are packed with explosives, which are detonated from far away.

PROJECT

LOOKING AT ROCKS

THE best way to learn about rocks and minerals is to look closely at as many different types as you can find. Look at pebbles on the beach and the stones in your garden. You will find that they are not all the same. Collect a specimen of each different rock type and compare them with each other. Give each rock a number to identify it, and keep a record of where you found it and what you can see in each piece.

A magnifying glass will magnify your rocks and help you see details that cannot be easily seen with the naked eye. To find out how many different minerals there are in each of your specimens, look for different colors, shapes and hardness. Testing the properties of minerals, such as hardness, can help to identify what sort of rock it is. Ask an adult to take you to the nearest geological museum to compare it with the specimens there. Why not start your own museum at home or at school?

A closer look
Clean a rock with a stiff brush and water. Stand so that plenty of light shines on the rock, and experiment to find the correct distance from the hand lens to the rock.

chisel
geological hammer
safety glasses
magnifying glass
pencils
hard hat
backpack
pocketknife
notebook
water
camera
gloves
field guide
compass
bucket
collecting bags and labels
map, mark on the map the places where you found your best specimens
newspapers, for wrapping your specimens in

Rock collecting
Here is the essential equipment that you will need for collecting rocks. Wear protective clothing, and always take an adult with you when you are away from home. Safety glasses will protect your eyes from razor-sharp splinters when hammering rocks. Do not strike cliffs or quarry faces with your hammer, but stick to blocks that have already fallen. Always remember that cliffs can be very dangerous—a hard hat or helmet will protect your head from falling rocks. Do not be greedy when collecting. Rocks and minerals need protecting just as much as wildlife, and some sites are protected by law. If you are collecting on a beach with cliffs behind, be careful not to be cut off by the tide.

PROJECT

TESTING FOR HARDNESS

You will need: *several rock samples, bowl of water, nail brush, coin, glass jar, steel file, sandpaper.*

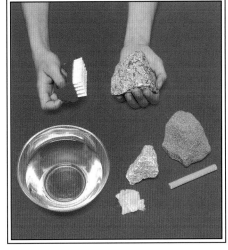

1 Clean some rock samples with water using a nail brush. Scratch the rocks together. On the Mohs scale, a mineral is harder than any minerals it can make scratches on.

2 A fingernail has a hardness of just over 2. Scratch each rock with a fingernail—if it scratches the rock, the minerals of which the rock is made have a hardness of 2 or less.

The hardest natural mineral is diamond, with a hardness of 10. It will scratch all other minerals.

3 Set aside those rocks scratched by a fingernail. Scratch those remaining with a coin. A coin has a hardness of about 3, so minerals it scratches are less than 3.

4 Scratch the remaining rocks on a glass jar. If any scratch the jar, then the minerals they contain must be harder than glass.

THE MOHS SCALE

- Mineral hardness is measured on a scale devised in 1822 by Friedrich Mohs. He listed ten common minerals running from 1, the softest, (talc) to 10, the hardest, (diamond).

5 Set aside any rocks that will not scratch the glass. Try scratching the remainder with a steel file (hardness 7) and finally with a sheet of sandpaper (hardness 8).

WHAT ARE MINERALS?

MINERALS are natural chemical substances that are present in all rocks. Most minerals are solid, but a few are liquid. Some minerals, such as sulfur and gold, are single elements. Others are made up of two or more elements. All rocks are a mixture of minerals. The igneous rock basalt, for example, which makes up most of Earth's oceanic crust, is a mixture of the minerals feldspar and pyroxene. Feldspar itself is a compound of oxygen, silicon and aluminum with various other elements. Silicates are the largest group of rock-forming minerals, all of which include silicon and oxygen. Quartz (the most common mineral in the Earth's crust) is a silicate. The minerals inside a rock usually form small crystal grains that are locked together to form a hard solid.

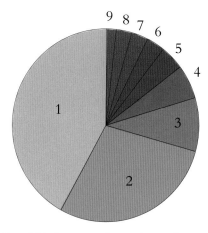

Just eight elements make up almost all minerals on Earth. Starting with the most common first, they are oxygen (1), silicon (2), aluminum (3), iron (4), calcium (5), sodium (6), potassium (7) and magnesium (8). All other elements make up (9).

Rock-forming minerals

Granite is one of the most common rocks found in the Earth's land crust. It is made mostly of quartz and feldspar with smaller amounts of mica and hornblende. As molten rock in the Earth's crust cools, the minerals form crystals and interlock with each other. Feldspar crystals are the first to crystallize and may be larger and more perfect in shape than the other minerals, which crystallize later. Feldspar is light (often pink) in color and quartz is gray and glassy. Mica is dark and silvery, while hornblende is usually jet black. Different granites have different amounts of each mineral, which is why granite varies in color from gray to reddish-pink.

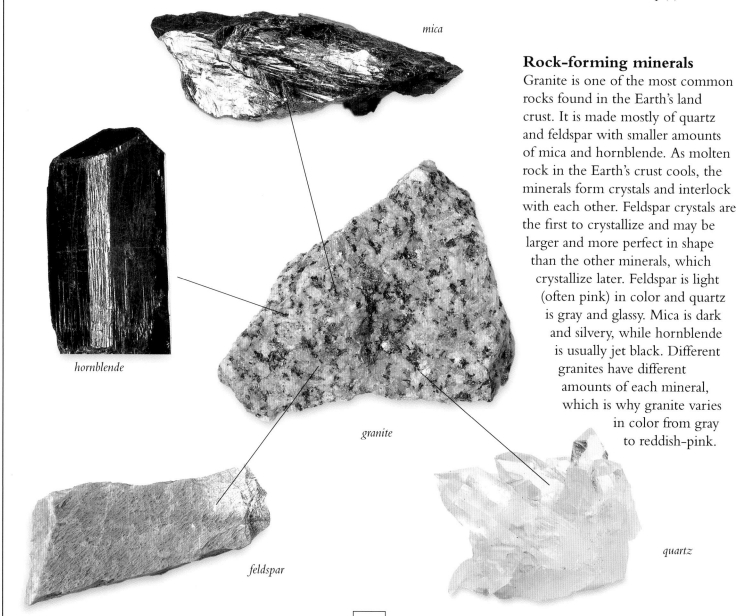

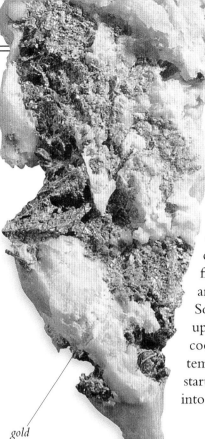
gold

Pure gold
A few minerals occur as single elements. A single element is one that is not combined with any other element. Gold is a good example of a single element. The mineral gold originally comes from hot rocks buried deep underground. Water flows through the hot rocks and dissolves the gold. Sometimes this water moves up toward the Earth's surface, cooling as it does. At lower temperatures the dissolved gold starts to harden and crystallizes into the solid form seen here.

FACT BOX
• The word diamond comes from the Greek word *adamas* (hardest metal).

• Ruby and sapphire are both rare forms of the mineral corundum, which is very hard. Tiny grains of corundum are used on metal nailfiles.

• Ore minerals contain metals, combined with the elements oxygen, sulfur and hydrogen.

emerald in mass of mica-schist

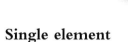

Single element
Diamonds grow under extreme pressure deep in the Earth and are carried to the surface in a rare volcanic rock known as kimberlite. The mineral diamond contains a single element, which is carbon.

Real versus synthetic
Gemstones such as diamonds and emeralds are rare and expensive. Today, the finest quality emeralds are found in the mountains of Colombia and Brazil, in South America.

synthetic emerald

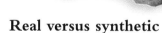

Colorful minerals
Under a microscope, a rock's crystals appear large enough to study. Scientists can identify the minerals by using filters that make polarized light. This gives each mineral its own range of colors.

Man-made crystals
Synthetic crystals can be made to grow in a particular size and shape, for a specific purpose. This is done by subjecting the crystals of more common minerals to carefully-controlled temperature and pressure. Some are used in the electronics industry for making computer chips. Others are manufactured for use in jewelry, such as the synthetic emerald cluster above.

CRYSTALS

pyrite

selenite

quartz

topaz

calcite

Most minerals found in rocks are in crystal form. They are highly prized for their beautiful colors and because they sparkle in the light. Crystals have often been associated with magic—the fortune teller's crystal ball was originally made from very large crystals of quartz. Most precious gemstones, including diamonds and rubies, are crystals.

Igneous rocks are usually made of interlocking crystals that form as hot magma (liquid rock) cools. The largest and best crystals are found in rock features known as veins. Veins are formed when hot, mineralized water rises up through the Earth. As the water cools, crystals form. Crystals may also grow when water on the Earth's surface evaporates. Each mineral variety forms crystals with a characteristic shape.

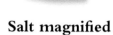

Salt magnified
These grains of ordinary table salt, seen under a powerful microscope, grew when salt water was evaporated. The grains are minute crystals, each containing billions of tiny particles known as atoms. The grains of salt are shaped like tiny cubes because of the way the atoms are arranged inside them. Each variety of mineral forms crystals with an individual shape.

Inside crystals
The shape of a crystal is controlled by the way that the atoms inside it are arranged. Imagine that the oranges in these boxes are atoms. In the left box the atoms are stacked in a disorderly way. Atoms that join like this do not produce crystals. Instead, they produce a material called glass. In a crystal, atoms join together in an orderly way, as in the box on the right.

Crystal faces
Crystals sparkle because their surfaces, or faces, reflect the light. Each individual mineral or group of minerals has faces that are always at the same angles relative to each other. There are seven main groups of crystals based on the arrangements of faces. The mineral crystals shown here illustrate five of the different groups.

The crystal lattice
The atoms in a crystal link together to form a three-dimensional framework known as the crystal lattice. This repeats itself in all directions as the crystal grows, giving the crystal its regular shape and controlling the angles between the faces. This picture shows ordinary salt or halite. The green balls represent chlorine atoms, the blue ones are sodium.

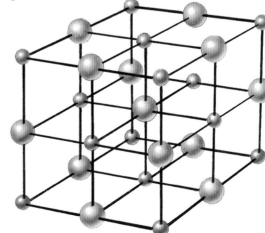

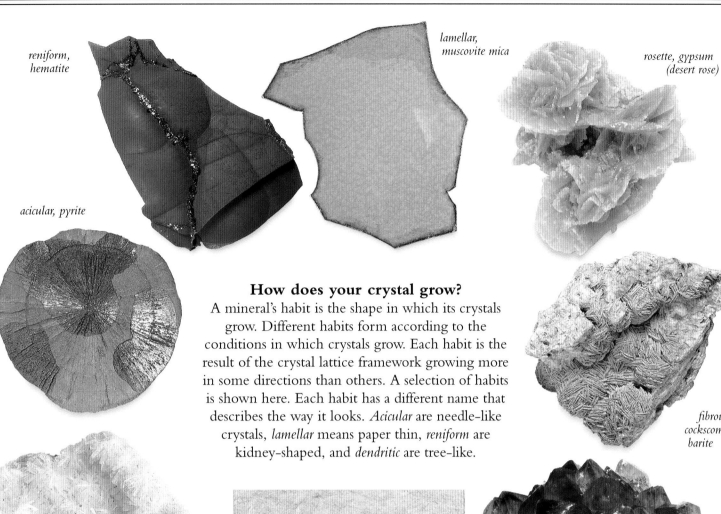

reniform, hematite

lamellar, muscovite mica

rosette, gypsum (desert rose)

acicular, pyrite

fibrous, cockscomb barite

How does your crystal grow?

A mineral's habit is the shape in which its crystals grow. Different habits form according to the conditions in which crystals grow. Each habit is the result of the crystal lattice framework growing more in some directions than others. A selection of habits is shown here. Each habit has a different name that describes the way it looks. *Acicular* are needle-like crystals, *lamellar* means paper thin, *reniform* are kidney-shaped, and *dendritic* are tree-like.

acicular, gypsum (daisy gypsum)

dendritic, manganese oxide

prismatic, amethyst

Crystal colors

The color of a crystal in natural light is a useful aid to identifying its mineral. Many types of mineral have characteristic colors, but several occur in a variety of colors. Quartz, for example, can be white, gray, red, purple, pink, yellow, green, brown, black and colorless. Citrine, rock crystal and rose quartz are three types of quartz. Amethyst is purple quartz.

rock crystal

citrine

rose quartz

Crystal twins

Sometimes crystals form so that two (or more) seem to intergrow symmetrically with each other. These are called twin crystals. Aragonite *(above)* often grows twinned crystals.

PROJECT

MAKING CRYSTALS

Most solid substances, including metals, consist of crystals. To see how crystals form, think what happens when sugar is put into hot water. The sugar dissolves to form a solution. If you take the water away again, the sugar molecules are left behind and join to reform into crystals. See this happen for yourself by trying the project below. Crystals can also form as a liquid cools. The type of crystals that form will depend on which substances are dissolved in the liquid. In a liquid, the atoms or molecules are loosely joined together. They can move around, which is why a liquid flows. As the liquid solidifies, the molecules do not move around so much and will join together, usually to form a crystal. You can see this if you put a drop of water on a mirror and leave it in a freezer overnight. Finally, make a simple goniometer (a device used to measure the angles between the faces of some objects).

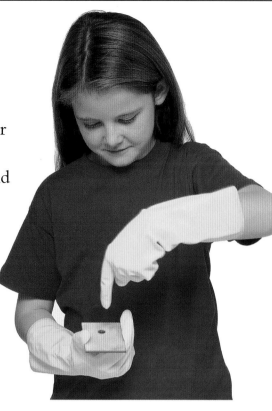

Ice crystals
A drop of water placed on a dry mirror will spread out a little, then freeze solid in the freezer. Examine the crystals that form with a magnifying glass.

You will need: water, measuring cup, saucepan, sugar, tablespoon, wooden spoon, glass jar.

GROWING CRYSTALS

1. Ask an adult to heat half a quart of water in a saucepan until it is hot, but not boiling. Using a tablespoon, add sugar to the hot water until it will no longer dissolve.

2. Stir the solution well, then let it cool. When it is quite cold, pour the solution from the pan into a glass jar and put it somewhere where it will not be disturbed.

3. After a few days or weeks, the sugar in the solution will gradually begin to form crystals. The longer you leave it, the larger your crystals will grow.

PROJECT

MAKING A GONIOMETER

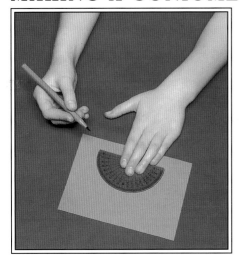

1. Firmly hold a protractor on a piece of card stock. Draw carefully around the protractor onto the card stock using a dark felt-tip pen or soft pencil. Do not move the protractor.

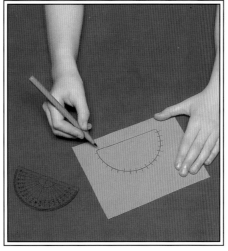

2. With the protractor still in place, mark off 10-degree divisions around the edge. Remove the protractor and then mark the divisions inside the semicircle.

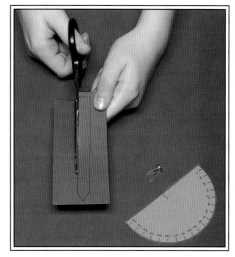

3. Cut out the semicircle. Now cut a thin strip of card stock about one inch longer than the base of your semicircle. Cut one end square and cut the other end into a point.

You will need: *protractor, two pieces of card stock, felt-tipped pen or pencil, scissors, ruler, paper fastener.*

Measuring the angles

People who study crystals sometimes use a device called a goniometer. It measures the angles between the faces of a crystal. The angle can help to identify a mineral.

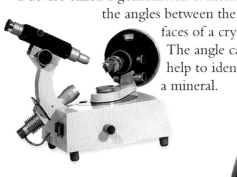

a simple goniometer

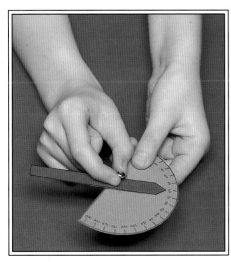

4. Make holes in the pointer and semicircle big enough for the paper fastener, and attach them with the fastener as shown. Flatten out the fastener on the back.

5. Collect some objects with straight sides or faces. Rest the straight face of your semicircle on one face of the object. Move the blunt end of your pointer onto the next face. The other end will point on the scale to the angle between the faces. Real goniometers are more complex and accurate than this, but they measure angles in a similar way to your homemade one and help to identify minerals.

IGNEOUS ROCKS

IGNEOUS rocks start off deep within the Earth as magma (molten rock). The name igneous means "of fire." The magma rises toward the surface where it may erupt as lava from a volcano, or cool and solidify within the Earth's crust. Igneous rocks that extrude, or push out, above ground are called extrusive. Those that solidify underground are called intrusive. Igneous rocks are a mass of interlocking crystals, which makes them very strong and ideal as building stones.

The size of the crystals depends on how quickly the magma cooled. Lavas cool quickly and contain very small crystals. Intrusive rocks cool much more slowly and have much larger crystals. The most common kind of lava, basalt, makes up most of the Earth's oceanic crust. Granite is a common intrusive igneous rock. It forms huge plugs, up to several miles thick and just as wide, in the continental crust. These are called plutons. They are often found under high mountains such as the Alps or the Himalayas.

Fine-grained granite
The magma that makes granite below the ground can also erupt at the surface. This Stone Age ax is made of a lava called rhyolite, and has razor-sharp edges. Its crystals are tiny.

Microscopic view
Geologists look at rocks under a special kind of microscope that shows the minerals in the rock. This is what an igneous rock called dolerite looks like. Notice the way the crystals fit together with no spaces between. Dolerite is a volcanic lava that formed beneath the ground. It has larger crystals than extrusive basalt.

Coarse-grained granite
This sample of granite is typical of igneous rocks. The large crystals give the rock a grainy texture. The crystals are large because they grew slowly as the liquid magma cooled down slowly.

Fine-grained basalt
Basalt is the most common extrusive igneous rock, especially in the oceanic crust. Basalt cools much more quickly than granite, so the crystals are smaller and the rock looks and feels smoother.

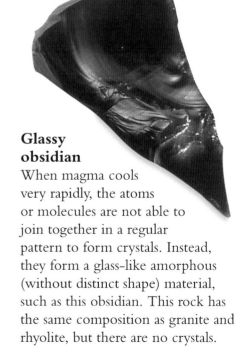

Glassy obsidian
When magma cools very rapidly, the atoms or molecules are not able to join together in a regular pattern to form crystals. Instead, they form a glass-like amorphous (without distinct shape) material, such as this obsidian. This rock has the same composition as granite and rhyolite, but there are no crystals.

Half dome
Millions of years ago, a huge dome of magma intruded under what is now Yosemite National Park in California. It slowly cooled to form granite. Over a period of time, the rocks around it were worn away by glaciers, exposing these dome-shaped hills of granite behind.

Granite tors
On Dartmoor in the southwest of England are the remains of large granite plutons that solidified below a chain of mountains. The mountains were eroded away, but the granite, being hard and weather-resistant, remains to form shapes known as tors. They look like huge boulders stacked on top of each other.

Rivers of liquid basalt
The islands of Hawaii are the exposed tops of huge piles of basalt that is still erupting after many thousands of years. The hot lava glows red in the dark and is capable of flowing for several miles before solidifying. Parts of Scotland and Ireland looked like this about 50 million years ago.

As the lava of the Giant's Causeway cooled, it cracked into interlocking, six-sided columns of basalt rock.

Giant's Causeway
The impressive columns of the Giant's Causeway in Northern Ireland are solid basalt. As lava reached the surface, it flowed into the sea, where it cooled and split into mainly hexagonal (six-sided) columns. The minerals that make up basalt, such as feldspars, pyroxenes and olivine, typically give the rock a dark-gray to black color.

PROJECT

MAKING IGNEOUS ROCKS

THE projects on these pages will show you how igneous rocks can be grainy and made of large crystals, or smooth and glassy. You will be melting sugar and then letting it solidify. Sugar melts at a low enough temperature for you to experiment with it safely at home. To make real magma, you would need to heat pieces of rock up to around 1,832°F until it melted! Even with sugar, the temperature must be high, so ask an adult to help you while you are carrying out these projects. You can also make the sugar mixture into bubbly "honeycomb" candy, which is similar in form to the rock pumice. It is like pumice because hundreds of tiny bubbles are captured inside the hot sugar.

"honeycomb" candy (pumice)

toffee (obsidian)

fudge (granite)

You will need: *sugar, water, saucepan, safety glasses, wooden spoon, milk.*

Fudge's grainy texture is similar to granite. Like obsidian, glassy toffee cools too rapidly to form crystals. The bubbles in "honeycomb" candy are like pumice.

MAKING CRYSTALLINE ROCK

1. Ask an adult to heat about 16 oz (1 lb) of sugar with a little water in a pan. Continue heating until the mixture turns brown, but not black, then add a dash of milk.

2. Let the mixture in the pan cool to room temperature. After an hour, you should see tiny crystal grains in the fudge mixture. Once it is completely cool, feel its texture.

PROJECT

MAKING GLASS AND BUBBLES

1. Use waxed paper to spread the butter on a metal baking sheet. Put in the freezer for at least an hour to get cold. Use oven mitts to take the sheet from the freezer.

2. Ask an adult to heat about 16 oz (1 lb) of sugar with a little water in a saucepan. The sugar dissolves in the water, but the water soon evaporates, leaving only sugar.

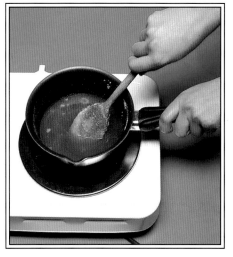

3. Stir the sugar mixture with a wooden spoon while it is heating. Make sure that the sugar does not burn and turn black. It should be golden brown.

You will need: *waxed paper, butter, baking sheet, oven mitts, sugar, water, saucepan, wooden spoon, safety glasses, baking soda.*

Baking soda gives off the gas carbon dioxide to form tiny bubbles.

4. Pour the mixture onto the cool baking sheet. After 10 minutes, the glassy and brittle toffee will be cool enough to pick up.

5. To make "honeycomb" candy, stir in a spoonful of baking soda in Step 3, just before you pour the sugar onto the sheet. This will make tiny bubbles of gas in your "magma."

SEDIMENTARY ROCKS

MANY of the most familiar rocks that we see around us are sedimentary rocks. Particles of rock, minerals and the shells and bones of sea creatures, settle in layers and then harden into rock over thousands of years. Rock particles form when other rocks are eroded (worn down) by the weather and are carried away by wind, rivers or ice sheets. They become sediments when they are dumped and settle. Sediments may collect in areas such as river deltas, lakes and the sea. Very large particles make conglomerates (large pebbles cemented together), medium-sized ones make sandstones and very fine particles make clays. Some sediments are made entirely of seashells. Others form when water evaporates to form a deposit called evaporite. Rock salt is a sedimentary rock and is used to make table salt.

Sandstone monolith
Uluru (Ayers Rock) is a monolith (single block of stone) in central Australia. It is the remains of a vast sandstone formation that once covered the entire region.

Clay
The particles in clay are too fine even to see with a microscope. Clay absorbs water, which makes it pliable and useful for modeling.

Limestone
This is one of the most common sedimentary rocks. It forms in water and consists mainly of the mineral calcite. Rainwater will dissolve it.

Conglomerate
A conglomerate contains rounded pebbles cemented together by rock made of much smaller particles.

Chalk
Chalk is made from the skeletons of millions of tiny sea creatures. The white cliffs of Dover on the southern coast of England are chalk.

Sandstone
There are many types and colors of sandstone. Each different type is made of tiny grains joined together. The grains are usually quartz.

Red sandstone
The quartz grains in this rock are coated with the mineral hematite (iron oxide) to give the red color. The rock is from an ancient desert.

Old and new

This cliff, on the coast of Dorset, in southern England, is made of layers of hard limestone and soft mudstone. Both rocks were once at the bottom of a shallow tropical sea at a time when dinosaurs roamed on land. The sea was teeming with all kinds of creatures that were buried and became fossils. The cliff is falling (making it dangerous) and will make new sediment as it is broken up by the waves and carried into the sea.

Beach pebbles

To see how the particles in sedimentary rocks form, look at the different sizes of pebbles on a beach. The constant back and forth of the waves grinds the pebbles smaller and smaller. When they settle and the conditions are right, these particles will form sedimentary rock.

Meeting place

Many sedimentary rocks form at deltas, where a river meets the sea. The river's flow slows right down, so sediment can no longer be carried in the water and is deposited. This picture of a delta on the island of Madagascar was taken from a satellite.

Rock salt pillars

The salt that forms rock salt is a chemical compound called sodium chloride, or halite. Rock salt forms as water evaporates from a salt solution, such as sea water. Here, pillars have formed in the extremely salty water of the Dead Sea, on the border between Israel and Jordan.

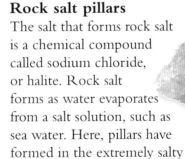

The red color of this halite is caused by impurities.

As water dries out, rock salt forms in pillars at the lake's edge.

FACT BOX

- Sedimentary rocks are said to be lithified, which means turned to stone. The word comes from the Greek word *lithos* (stone). The solid part of the Earth (including the crust) is called the lithosphere.

- Large areas of the world are covered with a yellow sedimentary rock called loess. The word comes from an old German word meaning loose. This is because loess consists of tiny dust particles that settled after being blown long distances by the wind.

ROCKS IN LAYERS

Strata sandwich
A multi-layer sandwich is like rock strata. The first layer is a slice of bread at the bottom. Each filling is laid on top with more slices of bread. When it is cut through you can see the many different layers.

SEDIMENTARY rocks form as small particles of rock accumulate at the bottom of seas and lakes, or in deserts. These particles settle to cover large areas and, over thousands or millions of years, new layers of sediment are laid down on top of existing ones.

As a result, most sedimentary rocks form in layers, called strata. The strata that are deepest underground are the oldest, because more recent layers are laid down on top of them. For this reason, sedimentary rock strata can provide valuable clues about the distant history of the Earth.

Once formed, sedimentary rocks may be subject to powerful forces caused by the movement of the Earth's crust. The forces squeeze the strata into folds and crack them. Along some large cracks, known as faults, blocks of rock slide past each other. Both folds and faults are often clearly visible in rock faces.

Folded strata
This rock face shows what happens when parts of the Earth's crust are pushed together in a collision zone. Geologists call downward folds synforms and upward folds antiforms. Folds are found in many sizes, from the microscopic to the gigantic. Mountain ranges have very large folds, sometimes with strata turned upside down. Old rocks may have been folded many times.

Layers of rock
Rock faces in cliffs, river valleys and mountains reveal sedimentary rock. Here you can see different layers of the same rock repeated. The layers are not of equal thickness. This suggests that conditions in this region changed many times in the past. Unequal thicknesses are common in layers of sediment found deposited at the mouths of large rivers.

Geological time chart		
Era	Period	Million years ago
Cenozoic	Quaternary	
	Holocene (epoch)	0.01
	Pleistocene (epoch)	2
	Tertiary	
	Pliocene (epoch)	5
	Miocene (epoch)	25
	Oligocene (epoch)	38
	Eocene (epoch)	55
	Paleocene (epoch)	65
Mesozoic	Cretaceous	144
	Jurassic	213
	Triassic	248
Paleozoic	Permian	286
	Carboniferous	360
	Devonian	408
	Silurian	438
	Ordovician	505
	Cambrian	590
Pre-Cambrian		4,600

Layers of seashells
In some sedimentary rocks there are layers made mostly of shells. In this picture the rock has split along the shell layer. It is like looking at the filling in a sandwich from above, after the top slice of bread has been removed.

Rocks and time
To put events in their correct place in time, it is necessary to have a framework or calendar that divides time in a way that is understood by everyone. Geologists have devised their own special calendar that is known as the Geological Timescale. This calendar starts 4,600 million years ago, which is the date of the oldest known rocks people have discovered. In this calendar, there are four major subdivisions, known as Eras. These are subdivided into Periods and Epochs (as shown in the chart above). So if you tell someone that you have found a Miocene fossil, he or she will know how old it is by referring to the calendar.

Clues to the past
Layers of sedimentary rock sometimes look different from each other. The differences are often evidence of climate changes in the past. Other differences might have been caused by the movement of tectonic plates.

Grand Canyon
The bottom strata of the 1-mile deep Grand Canyon, in Arizona is more than 2,000 million years old. Those at the top are about 60 million years old.

PROJECT

MAKING SEDIMENTARY ROCKS

To help you understand the processes by which sedimentary rocks are made and how they form distinct layers called strata, you can make your own sedimentary rocks. Different strata of rock are laid down by different types of sediment, so the first project involves making strata of your own, using various things found around the kitchen. The powerful forces that move parts of the Earth's crust often cause strata to fold, fault or just tilt and you can see this, too. In the second project, you can make a copy of a type of sedimentary rock called a conglomerate, in which small pebbles and sand become cemented into a finer material. Conglomerates in nature can be found in areas that were once under water.

The finished jar with its layers imitates real rock strata. Most sediments are laid down flat, but the forces that shape the land may tilt them, as here.

You will need:
large jar, modeling clay, spoon, flour, kidney beans, brown sugar, rice, lentils (or a similar variety of ingredients of different colors and textures).

YOUR OWN STRATA

1 Press one edge of a large jar into a piece of modeling clay, so that the jar sits at an angle. Slowly and carefully spoon a layer of flour about 1 inch thick into the jar.

2 Carefully add layers of kidney beans, brown sugar, rice, lentils and flour, building them up until they almost reach the top of the jar. Try to keep the side of the jar clean.

3 Remove the jar from the clay and stand it upright. The different colored and textured layers are like a section through a sequence of natural sedimentary rocks.

PROJECT

MAKING CONGLOMERATE ROCK

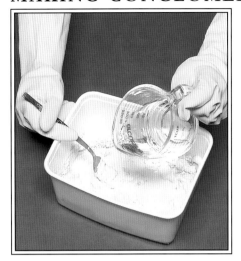

1. Put on a pair of rubber gloves. In a container, mix up some plaster of Paris with water, following the instructions on the package.

2. Mix some small pebbles, sand and soil into the plaster of Paris. Stir the mixture thoroughly to make sure they are all evenly distributed.

3. Let sit for 10 minutes, until the mixture starts to harden, then mold a small lump of it into a ball in your hand.

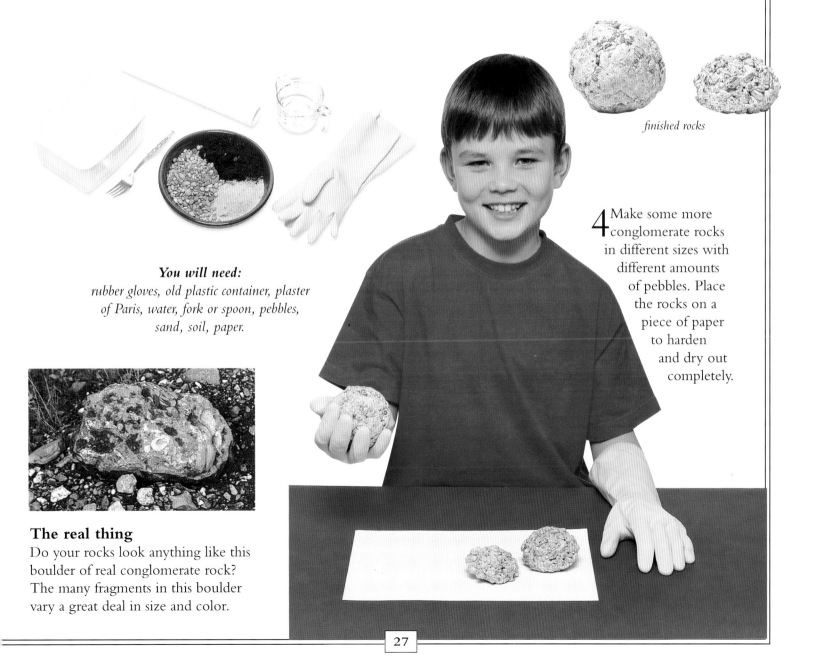

finished rocks

You will need:
rubber gloves, old plastic container, plaster of Paris, water, fork or spoon, pebbles, sand, soil, paper.

4. Make some more conglomerate rocks in different sizes with different amounts of pebbles. Place the rocks on a piece of paper to harden and dry out completely.

The real thing
Do your rocks look anything like this boulder of real conglomerate rock? The many fragments in this boulder vary a great deal in size and color.

27

METAMORPHIC ROCKS

THE word metamorphic means "changed," and that is exactly what these rocks are. Metamorphic rocks form when igneous or sedimentary rocks are subjected to high temperatures or are crushed by huge pressures underground. Such forces change the properties and the appearance of the rocks. For example, the sedimentary rock limestone becomes marble, which has a different texture and new minerals that are not found in the original limestone.

There are two types of metamorphism. In contact metamorphism, hot magma heats the surrounding rocks and changes them. In regional metamorphism, deeper rocks are changed when sections of the Earth's crust collide. In the intense heat and pressure of these collision zones the rocks start to melt in some places, new minerals appear and the layers are pushed into strange shapes.

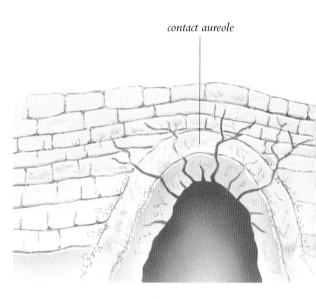

contact aureole

Contact metamorphism
This type of metamorphism occurs when an intrusion of magma bakes the rocks surrounding it. The zone of rock surrounding the magma is called a contact aureole. The magma alters the form and composition of the rocks. The changes are most noticeable close to the intrusion and gradually become less obvious farther away.

Mica crystals in slate give it a shiny, wet appearance.

The minerals form in bands, giving a foliated (layered) appearance.

The mineral olivine gives this marble its green color. The rest is mostly calcite.

Slate
This rock is formed from mudstone or shale, sedimentary rocks containing tiny particles of clay. Slate forms under very high pressure, but at a relatively low temperature. Because of this, fossils from the original rocks are often preserved but may be squeezed out of shape by the pressure.

Gneiss
Under very high temperature and pressure, many igneous or sedimentary rocks can become gneiss (pronounced "nice"). All gneisses are made of layers of minerals. In some, each layer is a different mineral. In others, they are different-sized crystals of the same mineral.

Marble
When heat and pressure alter limestone, which is a very common sedimentary rock, marble forms. Impurities in the limestone give marble its many different colors, including red, yellow, brown, blue, gray and green, arranged in veins or patterns.

Regional metamorphism

Movements in the Earth's crust can cause rocks to change in many different ways. The nature of the changes depends on the intensity of the pressure and the degree of heat. Slate or schist form at high pressures and low temperatures. For gneiss to form, both temperature and pressure must be very high.

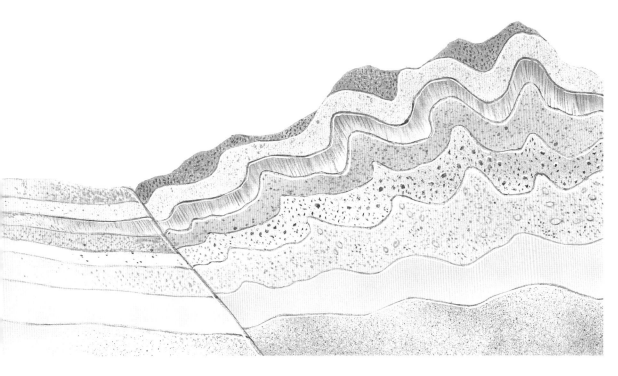

New minerals

Metamorphism causes new minerals to develop in old rocks. The large crystals of garnet in this schist were not in the original rock. The type of minerals that grow depends on the composition of the original rock and the strength of the metamorphism.

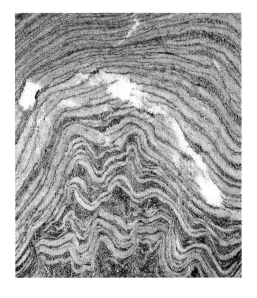

Squeezed almost to melting

When rocks are squeezed and heated, to temperatures close to their melting points, their consistency changes. They are no longer hard, brittle solids that can be cracked—they become plastic (easily molded). This gneiss was deformed when it was in a plastic state.

Why slate splits

The slate from this Welsh quarry will be used mainly for roof tiles. Slate splits naturally into thin slices along lines within its structure. This is because it has been squeezed. Then, small, lamellar (flattened), lined-up crystals grew in layers at right angles to the direction of the squeeze.

GEOLOGISTS AT WORK

GEOLOGY is the study of the history of the Earth, as revealed the rocks found in the Earth's crust. Scientists who study geology are called geologists. Some geologists specialize in certain branches of geology. For example, paleontologists study fossils, mineralogists specialize in identifying minerals and petrologists study the internal structure and composition of rocks.

Most geologists spend some of their time in the field, collecting samples and measuring various features of the rocks that they can see in outcrops. The rest of their time is spent in laboratories, analyzing the samples they have collected and the measurements they have made, often using a computer. Geologists keep detailed notebooks and make records of everything they discover about the rocks in the area where they are working.

The most important record is a geological map. This shows where different rocks may be found, how old they are and whether they are flat-lying, tilted or folded. This information also gives geologists clues about what is going on underneath the Earth's crust. The different rock types are shown on a map in different colors.

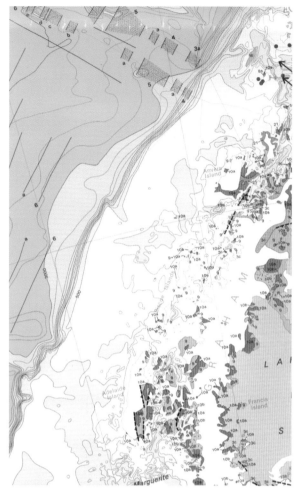

Planning ahead
Geological maps show the types and ages of rocks and how they lie on the land. From such information, geologists can work out where to look for minerals and underground water, and architects decide a suitable place for a building.

Fieldwork
These geologists are sampling water from a volcanic sulfur spring. Most geologists carry out fieldwork like this, collecting various kinds of sample for further study. In this case they will analyze the water to see what minerals have dissolved in it.

Sampling gas
These geologists are collecting samples of gas given off by a volcano. They wear gas masks because some of the gases may be harmful to their health.

Geological hammer
Geologists use a hammer to collect samples of rock. Sometimes they take a photograph of the hammer resting on an outcrop of rock to show the direction and scale of the rock.

> FACT BOX
>
> • After rocks have been formed, the radioactive elements in some minerals change into other elements at a steady rate over thousands of years. By measuring the amounts of the new elements, geologists can figure out a rock's age.
>
> • Another way to tell the age of rocks is to look at the fossils they contain. Different plants and animals lived at different times in the past. Identifying fossils can tell you where in geological time a rock fits.

Microscopic photograph
A micrograph is a photograph taken through a microscope. This picture shows the grains in sandstone greatly magnified. Grain size and shape can help to identify sandstones.

Rocks in close-up
Slices of rock are examined under a petrological microscope. This has special filters to polarize light. The slices are cut thinly so that light passes through them. The appearance of minerals in polarized light helps geologists to identify them.

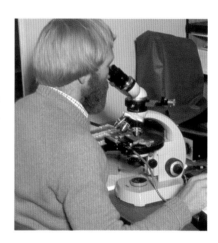

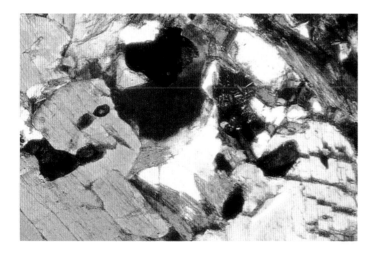

Polarized light micrograph
This micrograph of a thin section of a metamorphic schist was made using a polarizing microscope. It shows the crystals of different minerals that make up the rock. The micrograph helps to identify the rock's origin and the temperature and pressure at which it formed.

Density
These blocks are the same size and shape, but do not weigh the same. The materials they are made of have different densities. Density is used to identify minerals because samples of the same mineral will have the same density.

PROJECT

IDENTIFYING MINERALS

GEOLOGISTS use many different methods to identify the minerals that make up rocks. Each mineral possesses a unique set of identifying properties. Geologists use several tests to identify minerals, such as hardness (how easily a mineral scratches) and specific gravity (comparing a mineral's density to the density of water). They also look at streak (the color of a mineral's powder), luster (the way light reflects off the surface), transparency (whether light can pass through or not) and color (some minerals have a distinctive color in natural light). Dropping acid onto a sample to see if gas is given off is a simple test that can be carried out in the field. Try these simple versions of two tests that geologists use. They will help you identify some samples that you have collected. First of all, rubbing a rock on the back of a tile leaves a streak mark—the color of the streak can reveal the minerals that are present. Then you can calculate the specific gravity of a sample.

"Acid" test
Drop a rock into vinegar. If gas bubbles form, then it contains minerals called carbonates (such as calcite).

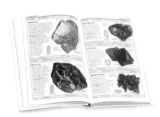

You will need:
white tile, several samples of different rocks or minerals, field guide.

STREAK TEST

1 Place a tile face down, so that the rough side is facing upward. Choose one of your samples and rub it against the tile. You should see a streak of color appear on the tile.

2 Make streaks using the other samples and compare the colors. Rocks made of several minerals may leave several colored streaks.

PROJECT

SPECIFIC GRAVITY TEST

1. Choose a rock and weigh it as accurately as you can to find its mass (weight). The figure should be in grams. Make a note of the mass.

2. Fill a clear measuring cup to the 1 cup mark with water. Now carefully place the first rock sample into the water.

3. Look carefully at the scale on the cup to read the new water level. Make a note of the level of the water in your notebook.

You will need:
mineral or rock samples, accurate scale, notebook, pen or pencil, measuring cup, water.

4. Take the figure you wrote down in Step 3 and subtract 200. This is the sample's volume in milliliters. Now divide the mass (weight) by the volume to find the density. You can use a calculator to do this sum if you want to.

The mass of a sample divided by its volume gives you its density, or specific gravity.

pyrite

beryl

Denser

If a mineral has a specific gravity (SG) of 5, it is five times as dense as water. Pyrite has a SG of 5 and beryl has a SG of 2.6. The atoms in pyrite are more closely packed together, making it denser.

READING THE ROCKS

When continents collide
An outcrop of gneiss, like the one above at the surface of the Earth, would tell geologists that the gneiss was once buried beneath high mountains like those on the right. Mountain ranges often develop when moving sections of the Earth's crust have collided, subjecting the rocks to the kind of strong squeezing and heating that creates gneiss. In time, the mountains are eroded away and the gneiss is exposed at the surface. The mountains shown on the right are the Alps, the result of Africa colliding with Europe.

GEOLOGISTS are interested in what has happened on planet Earth from the time it appeared over 4,000 million years ago to the present day. By reading clues in the rocks they can piece together how the climate changed in the past, how continents moved and how oceans and ice-sheets appeared and disappeared. No single geologist can discover all of this on his or her own. Each individual adds clues that are shared and considered by others in an endless process of detective work.

The first clues come from a careful study of exposed outcrops in the field. What minerals can be seen, and are they crystals? Are the rocks in layers? Are there fossils in them? Are the layers flat or tilted, or bent into shapes like waves? These are just a few of the questions that geologists ask.

More clues come from samples tested in the laboratory. What elements are in the samples? Are the rocks magnetic? How old are they? What kind of fossils do they contain? Each fragment of information is written down or stored on a computer. A store of knowledge about rocks is gradually assembled for geologists, so that they can draw conclusions about Earth's history.

Tropical sea
Limestones like this (left) begin in the sea. Sea creatures play an important part in their development. If you could travel back in time you would find the sea full of strange fish, shells and corals—similar to those in the picture on the right. When these kinds of creatures died, fragments of their shells were then consolidated into limestone.

Strong water
The large, rounded pebbles in this conglomerate outcrop (left) indicate that it came from water with currents strong enough to carry the pebbles. Such sediment could have started on a shoreline like the one on the right, or in a large, fast-flowing river.

Sedimentary basins
On the left, layers of hard sandstone alternate with thinner layers of soft shale, a rock rich in clay. Such sediments were washed into a shallow sea (right) to make what is called a "sedimentary basin." The changes in layers of sediment are caused by changes in the slope of the seabed, the position of the shoreline and the depth of the water.

SEEING INSIDE THE EARTH

How do geologists find out what is deep inside the Earth? Geologists can take measurements at rocky outcrops on land, which provide information about the rocks below. However, only a small proportion of the surface rock is exposed and accessible. Alternatively, geologists may probe down into the Earth. Information obtained from holes made during mining, or from natural caves in limestone areas, shows that the temperature of the Earth increases with depth. Smaller and deeper holes can be made by drilling. Modern drilling machines operate on land or at sea and can collect rock samples from several miles down. They are also able to drill sideways from the bottom of a vertical hole, and this increases the amount of information obtained about rocks at deep levels.

Another way of seeing inside the Earth is provided by geophysics, a sister science to geology. Geophysics is concerned with measuring things in rocks such as magnetism, gravity, radioactivity and the way in which rocks conduct electricity and sound. Three-dimensional maps can be made from measurements that show ups and downs of land. These maps show the distribution and formation of underground rocks.

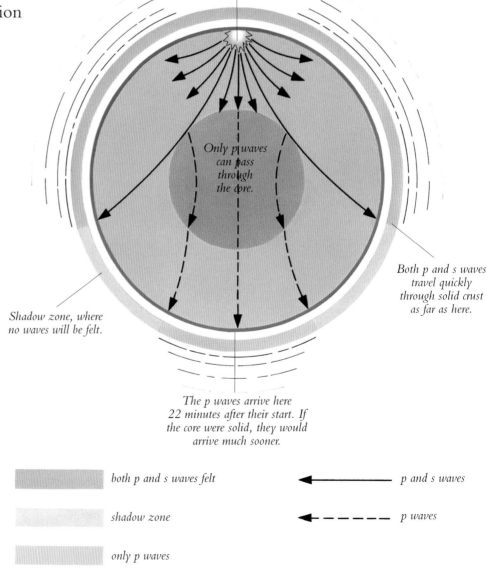

Earthquake starts here and makes vibrations of two kinds—p waves and s waves. The Earth vibrates like a bell, as the waves move outward in all directions.

Only p waves can pass through the core.

Both p and s waves travel quickly through solid crust as far as here.

Shadow zone, where no waves will be felt.

The p waves arrive here 22 minutes after their start. If the core were solid, they would arrive much sooner.

both p and s waves felt — p and s waves

shadow zone — p waves

only p waves

Vibrating Earth

Earthquakes have helped geologists to understand what the Earth is made of. When they occur it is as though the Earth has been hit by an enormous hammer. The Earth vibrates like a ringing bell. Waves of vibration move right through the Earth. They may be detected on the surface at places far away from the earthquake—even on the opposite side of the Earth. Very sensitive instruments called seismographs are used to measure the vibrations of an earthquake. These have been placed all over the world and are continuously switched on, ready to detect the next earthquake wherever it starts.

Earthquake waves travel at different speeds in different materials and will arrive at the seismographs at different times. All of the seismograph records ever taken show that the core of the Earth is mostly liquid.

Reflecting sound waves

Bats avoid flying into objects by detecting sound that is reflected from them. Something similar is used to look inside the Earth in seismic reflection surveys. These are commonly done from ships specially designed for the purpose. Loud bangs are made every few minutes by a compressed-air gun towed behind the ship. This sends sound waves down through the water into the rock below. Whenever the sound waves reach a different rock layer, they are reflected back to the surface. The reflections are picked up by a long string of underwater microphones, called hydrophones, also towed behind the ship. Powerful computers analyze the information to read the rocks beneath the seabed.

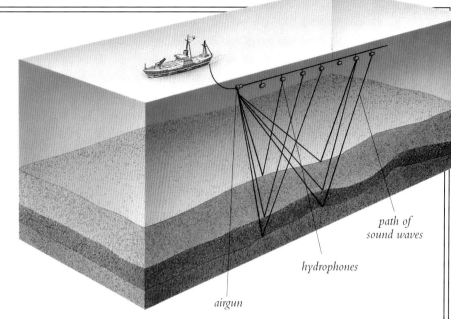

Seismic profile

One of the first things to be produced in a seismic reflection survey is a computerized drawing like the one on the left, showing the reflecting layers. From this, a geologist tries to figure out what rocks are present.

Drilling

Geophysical surveys provide many clues about what lies beneath the ground but they cannot identify rocks. Eventually holes have to be drilled and pieces of rock collected. The holes are made by turning a drill bit, studded with diamonds and attached to the end of a long string of steel pipes. More pipes are added as the bit goes deeper. The samples come to the surface either as broken chips or as solid cylinders of rock, known as core. These men are extracting core from a hole drilled on land.

Handling the data

A modern geophysical survey requires lots of computing power to process information about the inner Earth. This is the computer room in a modern survey ship making a seismic reflection survey. Computers control the airguns on the ship and process the information from each of the geophones.

LIMESTONE LANDSCAPES

calcite

flint nodule

The main mineral in limestone is calcite, which dissolves easily. Lumps or layers of flint or chert (forms of quartz) are often found in limestones.

LIMESTONE is a very common sedimentary rock, made largely of a mineral called calcite, which contains the element calcium. Limestone comes in many forms, chalk being one of them. The dazzling white chalk cliffs of southern England are made of limestone.

Limestone forms landscapes unlike others, because calcite dissolves quite easily in rainwater. This causes gaping holes to appear at the surface, which may suddenly swallow rivers and divert them underground. Somewhere downstream, the river will usually come out at the surface again.

Underground, many limestones are riddled with caves formed by the action of the slightly acidic water. Caves may form huge interconnected systems, some of which are still unexplored. Where water drips from rocks inside caves, dissolved calcite is deposited (left behind), forming pillars called stalactites and stalagmites. A limestone rock called travertine forms in the same way, leaving a thick coating much like the kind that forms in a kettle. An extreme limestone landscape called karst occurs where rainfall is fairly high. This is characterized by steep-sided limestone pinnacles separated by deep gullies. China has particularly spectacular karst scenery.

Karst scenery
These pinnacles are in Australia's Blue Mountains. Limestone areas create spectacular landscapes, called karst. Rainwater runs through cracks in the limestone to form underground caves and large holes called sink holes. Where the strata are tilted, deep cracks create pinnacles.

Limestone cones
Cones of limestone rock rise from beside the Li Jiang river near Guilin, China. The strange and beautiful towers were formed by intense downward erosion by rain and river water. The rainfall in this region is very high.

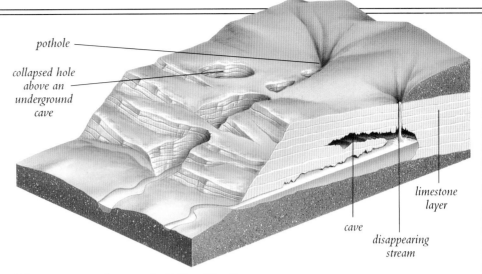

Travertine
This famous landscape is the Pamukkale Falls in Turkey. Over thousands of years, water from hot springs in limestone regions has deposited travertine (which is of the same composition as limestone) in beautifully shaped terraces. In some places travertine is quarried and used as a decorative building stone.

Limestone, the rock full of holes
Rainwater contains acid that dissolves limestone. As rainwater flows over the surface and through cracks in limestone areas, it slowly dissolves the rock. It leaves behind holes of all shapes and sizes, including potholes and caves. The holes have attracted many people who love to explore them. They are called speleologists (from *spelaion*, the Greek word for cave).

FACT BOX

- Karst scenery covers as much as 15 percent of the Earth's land area. The word karst was taken from the name of a region on the Dalmatian coast of Croatia, on the Adriatic Sea.

- In regions where the main type of rock is limestone, calcite dissolves into the water supply, forming hard water. In hot water pipes and kettles, some of the mineral is deposited to form brown, fur-like crystals.

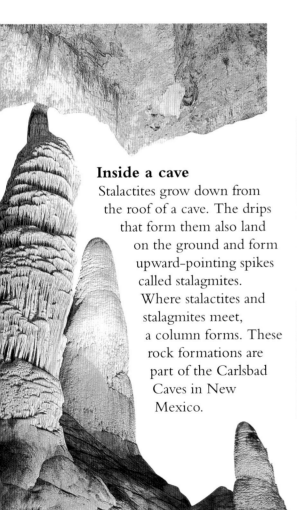

Inside a cave
Stalactites grow down from the roof of a cave. The drips that form them also land on the ground and form upward-pointing spikes called stalagmites. Where stalactites and stalagmites meet, a column forms. These rock formations are part of the Carlsbad Caves in New Mexico.

Limestone pavement
In many limestone regions, pavements with deeply furrowed surfaces are formed. The furrows, known as grikes, are made by water seeping through cracks and dissolving the limestone. The blocks of limestone formed by the furrows are called clints.

ROCK, WEATHER AND SOIL

Black sand
Most sand grains pass through soil at some stage in their recycling process. The sand on this beach is made mainly of the black mineral magnetite. It is formed when basalt, a volcanic lava, is broken down through weathering. It then moves in rivers to the sea where it makes black beach sand.

Soil is another stage in nature's recycling program in which one kind of rock is slowly changed into another. Soil is formed from a mixture of mineral grains, pieces of rock and decayed vegetable matter such as leaves and plants. It forms, by a process known as weathering, when surface rocks are broken down by plant roots and dissolved by water. The action of burrowing animals and insects further helps the weathering process. Some minerals dissolve quickly. Others, such as quartz, are not dissolved but stay behind in the soil as stones.

The soil itself gradually erodes. Particles in it are blown or washed away. There are many different kinds of soil, whose characteristics depend on climate and the type of rock from which they are formed. In hot, wet climates the soils are bright red and thick. In dry or very cold climates the soils are thin or completely absent.

Peat bogs
The amount of plant material in soil can be very high. Peat soil is made up mostly of dead moss.

Sand dunes
In dry climates where deserts form, sand is blown by the wind to form hills called dunes. These red dunes are part of the Namib Desert in southwest Africa. The quartz grains in the sand were once part of the soil that covered the region when the climate was wetter.

Sea mists provide enough moisture for some plants to survive.

Barkhans are crescent-shaped dunes that are always moving.

The wind creates beach-like ripples on the desert floor.

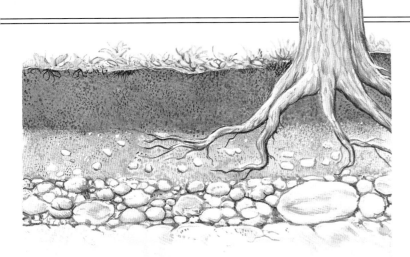

Soil in layers
Soil occurs in layers, known as horizons. There are four main horizons, called A, B, C and R. The A horizon (also called topsoil) is a layer of fine particles that supports the roots of plants and trees. In the B and C layers beneath the topsoil, the soil particles become larger. The R horizon is partly solid rock.

oak seedling

As trees grow, their roots help break down rock into soil. The roots work into cracks, splitting a rock apart.

Root tips grow down.

Dead leaves
As dead leaves and branches rot they release nutrients into the soil. Trees and other plants need these nutrients to grow.

Beach sand
These grains of sand have been enlarged under a microscope. The sand would have originated in soil—as small particles of rock. Over time, the grains have been transported to the sea.

Graded beds
Sand is moved along by flowing water. Where the current slows, sand and rock particles are deposited. This builds up layers of rock called a graded bed. Movements in the crust can tilt the bed at different angles.

River sand
The particles found in river sand, often of quartz, are generally bigger than in desert or beach sand. They are also more angular in shape. The particles come to rest either on the flood plain of a river or at the river mouth in the delta, where they form gritstone, a coarse type of sandstone.

41

PROJECT

WHAT IS SOIL MADE OF?

SAND and soil are made of millions of very small particles. Sand is formed from many types of rock, by a process called attrition (grinding down). This usually happens after the grains have been released from a soil that formed in a different place at a different time. Desert sand forms by attrition, as wind-blown mineral grains rub against each other. You can see how attrition forms small particles simply by shaking some sugar cubes together in a glass jar.

Soil is a mixture of particles of minerals, along with dead plant and animal matter. In the first project, a sample of soil is examined, using a sieve to separate particles of different sizes. In the second project, you can find out how graded beds of sediment form in rivers, lakes and seas, as first large and then finer particles of sediment are deposited.

Sugar shaker
Shake some sugar cubes in a jar. The cubes knock together as you shake. After a while you will see many tiny grains of sugar. A similar process occurs in desert sands and on beaches as mineral grains knock against each other and become smaller.

You will need:
gloves, trowel, soil, sieve, paper, magnifying glass, notebook, pen or pencil.

WHAT IS IN SOIL?

1 Put on the gloves and place a trowel full of soil into the sieve. Shake the sieve over a piece of white paper for a minute or so.

2 Tap the side of the sieve gently to help separate the different parts of the soil. Are there particles that will not go through the sieve?

3 Use a magnifying glass to examine the soil particles that fall onto the paper. Are there any small creatures or mineral grains? Note what you see.

PROJECT

BIG OR SMALL?

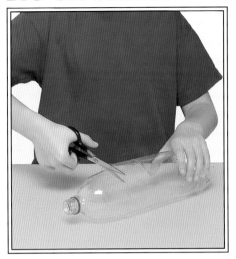

1. Using scissors, cut off the top of a large, clear plastic bottle. Throw away the top part.

2. Put small stones or gravel, soil and sand into the bottom of the bottle. Add water almost to the top.

You will need: scissors, large clear plastic bottle, small stones or gravel, soil, sand, wooden spoon, water.

3. Stir the stones, gravel, soil, sand and water vigorously. In a river, rock particles are combined and carried along by the moving water.

Sedimentary rocks often form in graded beds. This is because particles settle at different rates.

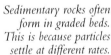

Humus

An important constituent of soil is humus. This is produced by animals called decomposers, such as worms and woodlice. These animals eat dead plant and animal matter, including leaves. As the matter passes through their bodies, it is broken down in their digestive systems.

floating humus and plant fragments

water made cloudy by very fine particles of clay

settled mineral particles

4. Let the mixture settle. You should find that the particles settle in different layers, with the heaviest particles at the bottom and the lightest on top.

PRESERVED IN STONE

coccosphere

radiolaria

foraminifera

THE remains of some organisms (plants and animals) that died long ago can be seen in sedimentary rock as fossils. After an organism dies, it may become buried in sediments. Slowly, over thousands of years, the sediments compact together to form sedimentary rock. The organic remains of the plants or animals disintegrate, but their shapes or outlines may remain. The hard parts of animals, such as bones and teeth, are preserved by minerals in the rock. Minerals can also replace and preserve the shape of the stem, leaves and flowers of a plant.

The study of fossils, called paleontology, tells us much about how life evolved, both in the sea and on the land. Fossils give clues to the type of environment in which an organism lived and can also help to date rocks. The fossil substance amber was formed from the sticky, sugary sap of trees similar to conifers, that died millions of years ago. The sap slowly hardened and became like stone. Insects attracted to the sweet sap sometimes became trapped in it, died there and so were preserved inside the amber.

Microfossils
Just as there are living organisms too small to see without a microscope (micro-organisms), there are also microfossils. These fossils are tiny marine organisms that lived during the Cretaceous period (about 65 to 144 million years ago). Millions of their remains are found in the sedimentary rock, chalk.

Early animals
Some fossils are the remains of animals that are now extinct. This trilobite, which is 600 million years old, is a distant relation of modern lobsters. Trilobites lived in ancient seas with all kinds of other creatures that are also now extinct.

HOW FOSSILS ARE FORMED

An animal or plant dies. Its body falls onto the sand at the bottom of the ocean or into mud on land. If it is buried quickly, then the body is protected from being eaten.

The soft parts of the body rot away, but the bones and teeth remain. After a long time the hard parts are replaced by minerals—usually calcite but sometimes pyrite or quartz.

After millions of years the rocks in which the fossils formed are eroded and exposed again. Some fossils look as fresh now as the day when the plant or animal was first buried.

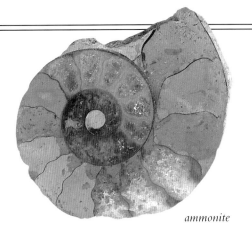
ammonite

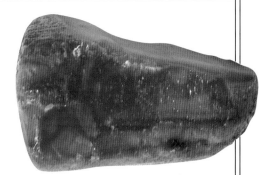
amber

Types of fossils

Five common fossils are shown here. Ammonites were hard-shelled sea creatures that lived between 60 and 400 million years ago. Fossils from sea creatures, such as shark's teeth, are often found, because their bones cannot decay completely underwater. The leaf imprint formed in mudstone around 250 million years ago and fern-like fossils are often found in coal. Amber is the fossilized sap of 60 million-year-old trees.

shark's tooth

leaf

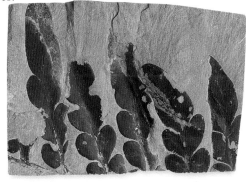
fern

FACT BOX

• Perhaps the most interesting fossilized animals are the dinosaurs, that lived between 65 and 245 million years ago. When their shapes are perfectly preserved, expert paleontologists can reconstruct the complete skeleton of the animal.

• Ammonites evolved (developed) rapidly and lived in many parts of the prehistoric world. Because geologists know how ammonites changed, they use the fossils to determine the ages of the rocks in which they are found.

Fossilized crab

This crab lived in the ocean around 150 million years ago. Marine limestones and mudstones contain the best-preserved fossils. Some limestones are made entirely of fossil shells.

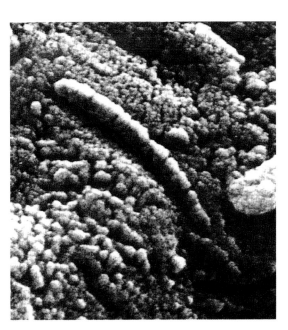

Martian fossils?

In 1996, scientists discovered what looked like tiny fossilized creatures in a rock that had originated on Mars. Everyone was excited about the possible proof that life had once existed on the planet, but it turned out that the marks were probably hardened mineral traces.

PROJECT
MAKING FOSSILS

THESE projects will help you to understand how two types of fossil came to exist. One type forms when a dead plant or animal leaves a space in the sedimentary rocks that settled around them. This is usually how the soft parts of an animal, or a delicate leaf, are preserved before they decay. The space in the rock is an imprint of the dead plant or animal. You can make a fossil of this kind using a shell, in the first project. In this case, the shell does not decay—you simply remove it from the plaster. In another kind of fossil, the spaces are formed when the decaying parts of an animal's body or skeleton are filled with minerals. This makes a solid fossil that is a copy of the original body part. Make this kind of fossil in the second project.

These are the finished results of the two projects. While you are making them, try to imagine how rocks form around real fossils. They are imprints of organisms that fell into mud millions of years ago.

You will need: *safety glasses, plastic container, plaster of Paris, water, fork, strip of paper, paper clip, modeling clay, shell, wooden board, hammer, chisel.*

These are a good example of dinosaur tracks. They were found in Cameroon, Africa.

MAKING A FOSSIL IMPRINT

1 In a container, mix up the plaster of Paris with water. Follow the instructions on the package. Make sure the mixture is fairly firm and not too runny.

2 Make a collar out of a strip of paper and the paper clip. Using modeling clay, make a base to fit inside the collar. Press in the shell. Surround the shell with plaster.

3 Let your plaster rock dry for at least half an hour. Crack open the rock and remove the shell. You will then see the imprint left behind after the shell has gone.

PROJECT

MAKING A SOLID FOSSIL

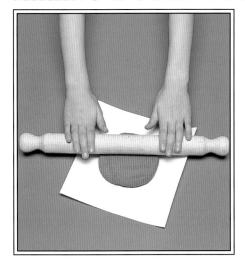

1. Put down a piece of paper to protect your work surface. Roll out a flat circle of modeling clay, about 1 inch thick.

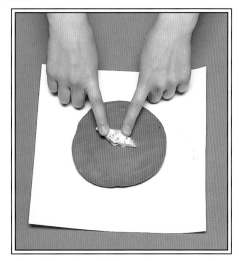

2. Press your shell or other object deep into the clay to leave a clear impression. Do not press it all the way to the paper at the bottom.

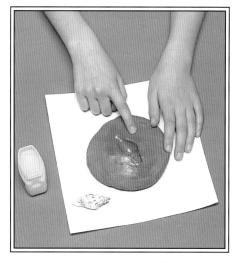

3. Remove the shell and lightly rub some petroleum jelly on the shell mold. This will help you to remove the plaster fossil later.

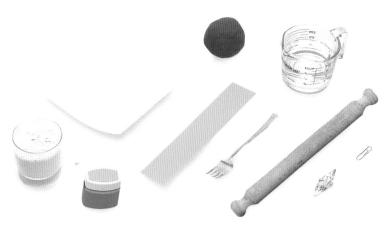

You will need: *paper, rolling pin, modeling clay, shell, petroleum jelly, paper clip, strip of paper, glass, plaster of Paris, water, fork.*

These jewel-like ammonite fossils were found in England at Lyme Regis, Dorset, a source of many different fossils.

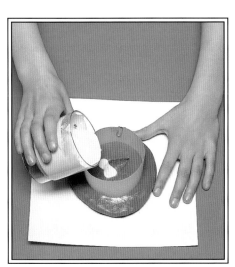

4. With the paper clip, make the paper strip into a collar for the mold. Mix up some plaster of Paris, pour it in and let set for half an hour.

5. Now carefully remove the solid plaster from your mold. In order not to damage them, paleontologists remove fossilized bones or teeth very carefully from rock or soil by cleaning them. They will do some cleaning in the field and the final cleaning back in the laboratory.

USEFUL ROCKS

All kinds of things are made from rocks and minerals. Early humans used hard rocks, such as flint, for stone tools. If you look around your house now, you will see many things that were once rocks or minerals. Bricks and roof tiles are made of clay. The cement that holds the bricks together is made from the sedimentary rock limestone. The plaster on the walls may be from a soft, powdery rock called gypsum. The glass in the windows comes mainly from quartz sand. China and pottery are made of clay. Laundry detergent powders and many plastic items come from the liquid mineral oil.

Plastics from oil
Plastics are made from an important group of minerals known as liquid hydrocarbons. One of these liquids, known as ethylene, is hardened through pressure and heat to form the solid plastic, polythene.

ax head

spear head

chopper

spear head

arrow head

scraper

Flint tools
Among the first tools made by humans were stone hand axes and blades. One common material for early tool-making was flint. It fractures easily to give a sharp edge and could be flaked to form many different tools.

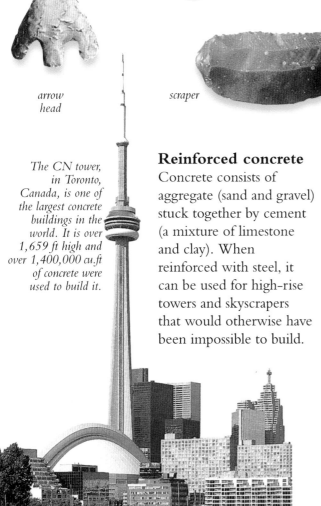

The CN tower, in Toronto, Canada, is one of the largest concrete buildings in the world. It is over 1,659 ft high and over 1,400,000 cu.ft of concrete were used to build it.

Reinforced concrete
Concrete consists of aggregate (sand and gravel) stuck together by cement (a mixture of limestone and clay). When reinforced with steel, it can be used for high-rise towers and skyscrapers that would otherwise have been impossible to build.

Crushed limestone
In many countries, limestone is used in larger amounts than any other rock. A layer of limestone, crushed into walnut-sized pieces, makes a perfect base for tarmac on major roads.

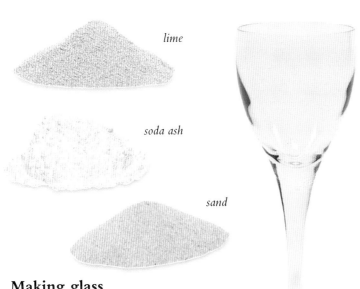

Making glass

High-quality glass is made by melting pure sand, at very high temperatures. Most ordinary glass is made from a mixture of sand with soda ash and lime, because it melts at a lower temperature. Glass was first made by the ancient Egyptians.

Porcelain china

The most highly prized material for making china is porcelain because it is strong and waterproof. Other ceramics, such as earthenware, absorb water more easily because they are made from coarser clays. True porcelain is made using very fine china clay, known as kaolin.

Colored pigments also made from minerals are used as decoration on porcelain.

Marble slabs

Polished slabs of marble make an eye-catching surface for floors and decorative objects. Marble is a metamorphic limestone, valued because it can be easily cut and polished. Some limestones and granites are popular for the same reasons.

Shaping clay

When clay is mixed with water, it becomes malleable (it can be shaped easily). Clay objects are first molded into shape, then baked in large kilns. Shiny objects are made by coating them with a glaze, which can be made in various colors. Unglazed pottery is called terra-cotta, which is from the Italian word for "baked earth."

FACT BOX

- In the mid-1700s, Coalbrookdale in England became one of the first industrial towns due to its interesting geology. It has an unusual sequence of rocks that includes layers of clay (for pottery and bricks), and coal, iron ore and limestone, which are the essential ingredients for iron-making. Running out of the rocks was natural bitumen, a material that was used to make machine oil. It was also used to waterproof the boats that transported the cast iron.

COAL

peat

lignite or brown coal

black coal

THE shiny black material that we call coal is a very useful material. It is called a fossil fuel, because it is from the fossils of dead plants. Burning coal in power stations is one of the ways in which electricity is generated. To most geologists, coal is a type of sedimentary rock made of solid minerals. These originally come from plants that died long ago and became rapidly buried by other sediments, commonly sandstones. Much of the world's coal formed from plants that lived and died in the Carboniferous period, which was between 286 and 360 million years ago. At that time, tropical rainforests existed across Europe, Asia and North America. Coal of a different age is found in India and Australia.

Lush rainforest
Damp, swampy rainforest is a similar environment to the one in which coal formed millions of years ago. To produce coal, the remains of plants lay submerged underwater, in swamps or shallow lakes, for millions of years.

anthracite

Types of coal
The hardest, best quality coal is called anthracite. More crumbly black coal and lower grade brown coal (lignite) do not give as much energy when burned. Lignite was formed more recently than black coal and anthracite. Peat is younger still. It forms even today.

Coal seams
Coal is found in layers known as seams. The largest may be several yards thick. Between the seams there are layers of sandstone or mudstone. This machine is extracting coals from a seam at the surface.

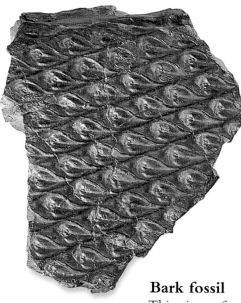

Bark fossil
This piece of coal clearly shows the bark patterns of the plant it came from. Plants such as tree-like horsetails, primitive conifers and giant ferns grew in huge forests.

Stoking the fire
Coal releases large amounts of energy as heat when it burns. A steam locomotive uses heat from burning coal to produce steam to drive its wheels. The driver keeps the fire well stoked.

Victorian jet brooch

Since the Bronze Age, jet has been carved into many decorative objects. It was popular in Victorian England for mourning jewelry, which was worn in memory of the dead.

Black as jet
Jet is a material that is similar to coal. It is formed from pieces of driftwood that settled in mud at the bottom of the ocean. It is very light and is sometimes polished and carved into intricate shapes.

natural jet

Digging out the coal
In the past, coal was dug from deep underground mines. Today it is mostly taken from large, open pits using gigantic ground-moving equipment. After the coal has been removed the holes left behind are turned into recreation areas. Lakes are created for boating and fishing, and old areas of waste rock are planted with grass and trees. Years ago, this restoration work was not always carried out, so that coal mining areas were unpleasant places to live.

At the coal face
A small amount of coal is still mined underground. Tunnels are built so that people and machines can reach an exposed seam of coal, called the coal face. In areas that are difficult to reach, the coal is mined by hand.

Coal-fired power station
Power stations are often built near coal fields. Coal is used to heat water to make steam. Steam-driven turbines run huge generators that produce electricity.

LIQUID MINERALS

MINERALS in rocks also occur in liquid form. Of all the minerals on Earth, one is used by you more than all the others—water. This liquid mineral has greatly influenced the way Earth developed. Without water there would be no oceans, no rivers or lakes, no rain clouds and no plants or animals. The atmosphere would be quite different. It would probably be made of carbon dioxide, similar to all the other dry, lifeless planets in the solar system.

Water is constantly being recycled, moving between the atmosphere, the oceans and the rocks of the Earth's crust. As it moves through the crust, it dissolves, grinds, freezes and thaws, changing the rocks it passes through.

In the oceans, living plants and animals have played a part in the making of another important group of minerals. These are liquid hydrocarbons, from which many essential products such as gasoline, lubricating oils and natural gas are made.

Water supplies
When you turn on a tap you may not think about where the cool, clean water comes from. It may be surface water that has come from a lake or river, or it may be groundwater that has come from deep inside the Earth. Both sources are stopping places in the great cyclic journey made by water. Occasionally water from different points in the water cycle is used. People living on small ocean islands may collect rainwater for their needs. In some countries where rainfall is low, seawater is turned into drinking water by removing the salt.

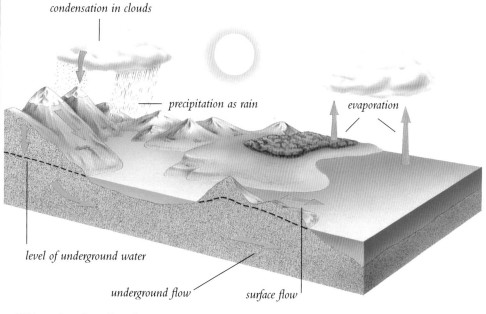

condensation in clouds
precipitation as rain
evaporation
level of underground water
underground flow
surface flow

Water's circular journey
As water moves through its cycle, it changes from water vapor in the atmosphere to drops of liquid rain. Some of the rain falls into rivers and streams that flow into the sea. This water is known as surface water. Some water percolates downward through cracks and pores in rock and becomes groundwater. Sometimes, groundwater seeps out at the surface in springs, but usually it has to be pumped up through wells or boreholes. Throughout its cycle, water slowly reacts with rock and dissolves elements from it. Two types of water are most commonly found, hard water and soft water. The first contains minerals, the second does not.

Clean water
Surface water, such as river water, is often cloudy because it contains particles of sediment such as clay. Water that has not been cleaned at a water plant also contains microscopic matter that can cause disease. Groundwater contains dissolved elements such as calcium and magnesium, but it is clear and sparkling because the rocks it has been filtered through have cleaned it.

The story of oil begins in warm seas full of living things. As they die, they fall to the sea floor to decay into thick black mud. Unusual events must then occur if oil is to form.

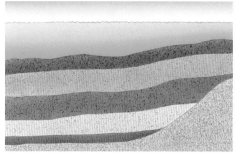
In time, the mud must be buried beneath many layers of sand with clay in between. The sediments must sink deeper and deeper and also become hotter.

After millions of more years, the sediments must come under pressure and fold. Oil from the black mud is then forced into the sandstones and trapped under layers of clay.

From plankton to oil

Liquid hydrocarbons, such as oil, are mixtures of the element carbon and the gas hydrogen. Geologists believe that carbon and hydrogen originally come from small sea creatures known as plankton. When the plankton die, their remains accumulate on the seabed. As they become buried in mud they decompose and make droplets of oil. With time, the mud is buried under very thick layers of sediment. In some situations, the droplets join together to make big underground pools. These get bigger and bigger until they form an oil reservoir.

Extracting oil

Oil reservoirs are often found beneath the sea. Rigs are built over the reservoir to pipe the oil and gas to shore. The rig shown above, in the North Sea, is built on legs that rest on the sea bed. The men who work on the rig are carried to and from shore by helicopter.

The hazards of producing oil

Oil in a natural reservoir is under pressure from whatever is above it. It is usually held in place by a layer of nonporous rock immediately above, which keeps it from seeping away. This is known as the cap rock. When an oil well is drilled through the cap rock into the reservoir, the pressure forces the oil into the well. Strong valves must be fitted to control the flow of oil and turned off when needed. In the well shown on the left, in Kuwait, there was an accident causing a jet of burning oil to rise high into the air. It was eventually extinguished, and the well was made safe again.

METALS FROM ROCKS

Among the most important minerals are metal ores. These contain minerals rich in metallic elements, such as iron, copper and tin. To extract the metal from its ore, the ore must first be separated from the rock in which it is found. The ore must then be heated with various other substances in a process known as smelting.

A few metals exist in rocks as a pure element. Gold is the best example and is found as veins and nuggets in many types of rock. Gold is dug from the rock in mines, although in some parts of the world it can be found as grains in river sand. The largest grains are known as nuggets.

Metals are useful to people because they last for a long time. They can also be shaped into many objects or drawn out into a fine wire that will conduct (let through) heat and electricity.

Reclaimed by nature
Metals in the Earth exist joined to other elements. Some quickly rejoin after extraction. Iron quickly rejoins with oxygen and water to form iron oxide, better known as rust.

Gold Rush
Sometimes rocks surrounding a nugget of gold are eroded by a river. Occasionally, the gold is released into the river and can be recovered through a hand-sieving process called panning. This picture shows the 1849 California Gold Rush, when thousands hoped to make a fortune from gold.

Metal sculpture
In many cities you can find statues made of metal. Most are made of bronze, a mixture of copper and tin. This famous statue of Eros in Piccadilly Circus, London, is unusual because it is made of aluminum, a silvery metal. At the time the statue was erected in 1893, aluminum was very difficult to make. It is not easily smelted like other metals and is extracted from the Earth by using electricity.

Fine gold wires
In this close-up view of a microprocessor, you can see very thin wires of gold. Most metals can be drawn into fine wires—a property called ductility. Gold is the most ductile of all metals. It also resists the corrosive effects of many chemicals, making it an important metal to the electronics industry and in dentistry.

FACT BOX

• A mixture of two or more metals is called an alloy. The first alloy people made was bronze (a mixture of copper and tin). Bronze is strong and durable and was used to make tools and weapons as long as 6,000 years ago.

• Gold is a rare and precious metal. In the hope of finding gold, some people have mistaken other minerals for it. These minerals are known as fool's gold and include shiny pyrite and brassy-colored chalcopyrite.

stainless steel cutlery

hematite (iron ore)

carbon

Copper

Water pipes and plastic-coated electrical wire are two important things made of copper. The metal is taken from its ore first by removing elements that are not copper. Then it is heated in a furnace with a blast of oxygen.

copper pipes

native copper

Iron

In iron ore, atoms of iron are joined to atoms of oxygen. To produce iron metal, iron ore is heated in a furnace with carbon. The carbon takes away the oxygen atoms, leaving the metal behind. Adding more carbon produces steel, which is harder and less likely to snap or corrode. Steel is the most widely used metallic substance.

aluminum foil

bauxite (aluminum ore)

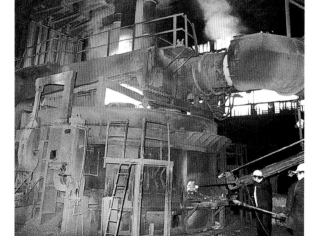

Aluminum

In aluminum ore, atoms of aluminum are tightly joined to atoms of oxygen. Powerful currents of electricity are used to separate the aluminum from the oxygen.

Smelting iron

Iron ore is heated in a blast furnace to produce almost pure iron. In a process known as smelting, the blast furnace blows hot air through to remove impurities from the ore.

ROCKS AND BUILDINGS

Rocks have been used for building ever since people discovered ways of cutting blocks of stone from the ground. Stonehenge, in southwest England, is evidence that about 4,500 years ago people knew how to cut, shape and move giant slabs of rock into position. Exactly how they did this is a mystery. In South America, the Incas were building large stone buildings long before the voyage of Christopher Columbus in 1492. All over the world, many examples of stone buildings can be seen. Over the last hundred years natural stone gave way to man-made brick and concrete, but today stone is again being used in buildings.

In the past, builders used whatever rock was common in the area, giving many towns and cities an individual character. For large buildings, the blocks of rock must be free of cracks and natural weakness, and splittable in any direction. They are then called freestone. The best freestones are igneous rocks, especially granite. Of the sedimentary rocks, some sandstones and limestones make good freestone, while marble is the best freestone among the metamorphic rocks. If sandstone splits easily along one direction into thick slabs, it may be used as a flagstone for floors or walls. Slate is a metamorphic rock that can be split into very thin sheets and is ideal for roofing.

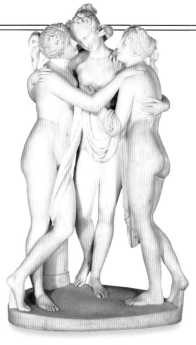

The Three Graces
This sculpture is by Antonio Canova, thought to be the greatest sculptor of the 1700s. It is made of marble from Italy and shows the detail that skilled sculptors can achieve using this stone. Polished marble has a special attraction because light is able to penetrate it and is then reflected back to the surface by deeper crystals.

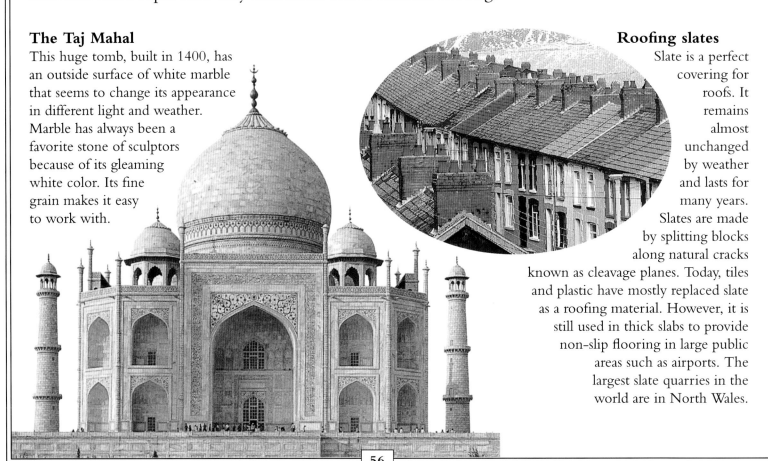

The Taj Mahal
This huge tomb, built in 1400, has an outside surface of white marble that seems to change its appearance in different light and weather. Marble has always been a favorite stone of sculptors because of its gleaming white color. Its fine grain makes it easy to work with.

Roofing slates
Slate is a perfect covering for roofs. It remains almost unchanged by weather and lasts for many years. Slates are made by splitting blocks along natural cracks known as cleavage planes. Today, tiles and plastic have mostly replaced slate as a roofing material. However, it is still used in thick slabs to provide non-slip flooring in large public areas such as airports. The largest slate quarries in the world are in North Wales.

England's earliest stone building
About 4,500 years ago a hard sandstone was used for the largest stones at Stonehenge in southwest England. Like the Incas, the builders managed to shape and lift these very large stones without using metal tools.

A perfect fit
These stones at the city of Machu Picchu in Peru are of white granite. They were shaped by the Inca people who did not have metal tools. Even now, no one knows how this was done. The blocks fit together so well that a knife cannot be pushed between them. They are even earthquake-proof. The Incas used stone available in the area, such as limestone, rhyolite and andesite.

Walls without cement
In many upland areas where trees do not grow well, stone is used for field boundary walls rather than hedges. They are made with whatever stone blocks can be found near by. Great skill is needed to interlock the different-sized blocks together, without the use of a binding substance, such as cement.

St. Paul's Cathedral
After the Great Fire of London destroyed a large area of the city in 1666, the architect Sir Christopher Wren was given the task of rebuilding St. Paul's Cathedral. It took him over 35 years. He chose to build the cathedral using a white limestone from Dorset, England, called Portland stone. Over 6 million tons were used in the new St. Paul's and other buildings in London. The stone was carried in barges along the coast and up the River Thames.

Stone working today
These huge blocks are of white metamorphic marble which is found in the Carrara Mountains of Italy. The blocks are cut from the quarry by drilling lines of closely spaced holes which are then filled with explosive. Carrara marble has a very fine-grained texture and a beautiful luster when it is polished.

ROCKS IN SPACE

chondrite
(stony) meteorite consisting mainly of the minerals olivine and pyroxene

iron meteorite
consisting of nickle-iron, which is strongly magnetic

shergottite
(stony) meteorite consisting of two different basalt rocks

We have evidence that rocks and minerals exist on other planets. Other planets and rocky material are in orbit around the Sun. They were formed when the solar system was created, some 4,600 million years ago. Some fragments of the rocky material exists as small particles called meteoroids. Every day, tons of this material hits the top of the Earth's atmosphere. Here, friction causes it to heat up and vaporize, sometimes causing a spectacular display called a meteor shower, or shooting stars. Larger particles do not vaporize completely, and a few actually hit the ground. These rocks are called meteorites.

Some meteorites are from the Moon or Mars. They were chunks of rock that were thrown off the planet when rock fragments from space bombarded the surface, forming craters. The surface of the Moon is littered with craters. When meteorite craters have formed on Earth, they have usually been covered over or destroyed by geological processes, such as the formation of mountains or erosion by the weather.

Meteorites
There are three main types of meteorite—stony, iron and stony-iron. Stony meteorites (the most common) are made of rock. The others contain nickel and iron.

Craters on the Moon
Many millions of years ago the Moon was constantly bombarded by meteorites that hit its surface, making huge craters. The Earth may once have looked like this. It has few craters today because of the action of wind and rain and movements of the crust that makes up the Earth's surface.

Meteor crater
This hole in the ground in Arizona is a crater. It was formed by the impact of a meteorite that fell about 25,000 years ago. About six craters of this size exist on earth. Most craters are covered up, filled with water or were eroded long ago.

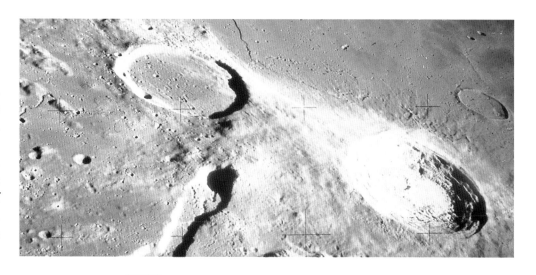

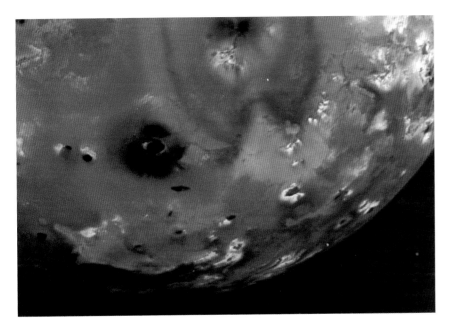

Rocks on Mars
The Pathfinder mission landed on Mars in July 1997. Aboard was a robot probe, called the *Sojourner Rover*, which studied rocks on the surface and sent back photographic images.

Jupiter's moon
This picture, taken in 1979 by the spacecraft *Voyager 1*, shows Io, one of Jupiter's moons. It clearly shows that other planets in the solar system have volcanoes. However, none of the planets is known to have an outer layer that moves over a liquid center, as on Earth.

The red planet
Mars is called the red planet because its surface is covered with red iron-oxide dust. It is the most Earth-like of the planets and may hold important clues for Earth's future climate. Four huge volcanoes and an enormous canyon scar its dry surface.

Asteroids
Many meteorites are thought to be broken fragments formed by the collisions between asteroids (small celestial bodies). Most asteroids orbit the Sun in a belt between Mars and Jupiter. Inside an asteroid is a central core of metal, which is surrounded by rock.

FACT BOX
- Astronauts have brought about 840 lbs of rock from the Moon to Earth. The most common type of rock on the Moon is basalt. It is the same as the basalt on Earth and is formed from solidified lava from volcanoes.

- Only one person is ever known to have been hit by a meteorite. It happened in 1954, in Sylacauga, Alabama. The person was not hurt because the meteorite had already bounced on the ground.

Shooting stars
Meteoroids (small objects from space that hit the Earth's outer atmosphere) travel at high speed. As they pass through our air, they heat up and glow yellow-white, appearing as a bright streak across the sky. This is called a meteor, or shooting star.

STONES FOR DECORATION

MINERALS that are highly prized for their beauty are called gemstones. The main use of gemstones is in jewelry or other decorative work, although some are also used in industry. Around 90 minerals are classed as gems. About 20 of these minerals are considered important gems, because of their rarity. These include diamonds, the most valuable of all gemstones. Some minerals provide more than one type of gem. For example, different types of the mineral beryl form emerald, aquamarine, heliodor and morganite. Gems such as ruby and emerald are distinctive because of their deep color. The different colors of gemstones are caused by metal impurities in the mineral. Other minerals, not necessarily gemstones, are prized for their color and ground down to pigments. These can be mixed with water to make paint.

Rock paintings
Aboriginal Australians first drew rock paintings like this one thousands of years ago. They used earthy-toned mineral pigments, such as umber (red-brown) and ocher (dark yellow).

raw umber *brown umber* *yellow ocher*

Pigments
Mineral colors have been used for thousands of years as pigments. Rocks containing colorful minerals are ground down, mixed with a binder, such as egg yolk, fat or oil, and used as paint.

The Millennium Star Diamond
This magnificent pear-shaped diamond from South Africa is one of the finest ever discovered. The hundreds of facets have been cut with great skill to bring out the perfection of the stone. It was faceted to mark the new millennium and was displayed in the Millennium Dome in London, England.

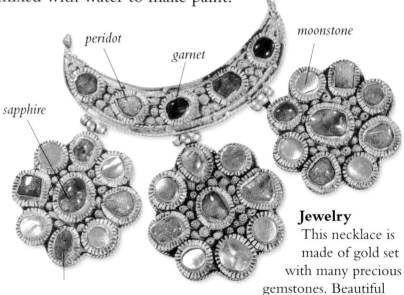

peridot *garnet* *moonstone* *sapphire* *pink sapphire*

Jewelry
This necklace is made of gold set with many precious gemstones. Beautiful minerals have been used for thousands of years in decorative jewelry.

Cameo
The gemstone agate occurs in layers of different colors. This makes it possible to carve in layers, using a decorative technique known as cameo. In a cameo, the top layer is carved to reveal the lower one as a background. This Greek cameo is of Alexander the Great.

Lapis lazuli
A mixture of the minerals lazurite, pyrite and calcite forms lapis lazuli. Its blue color is caused by the presence of sulfur in the mineral lazurite. The ancient Persians were the first to crush the rock and use the pigment for ultramarine paint.

Specks of pyrite in lapis look like gold.

natural lapis lazuli

Amber beads
Amber is called an organic gem because it is formed from prehistoric tree sap. Most gem-quality amber comes from the shores of the Baltic Sea.

Jade carving
The characteristic green color of jade comes from atoms of iron metal. Two different minerals are called jade, jadeite and the more common nephrite.

natural nephrite

Blue john is cut and polished to show off the lacy banding.

ruby   *diamond*

Blue john
The distinctive purple banding in the mineral fluorite is commonly known as blue john. In fact, the bands vary in color from purple-blue to yellow. Fluorite is a common mineral and is found in limestones.

natural blue john

Blood and fire
A diamond's fiery brilliance, hardness, purity and rarity make it the most valuable gem. Rubies are rare forms of the mineral corundum. Their blood-red color comes from the metal chromium.

PROJECT

USING ROCKS AND MINERALS

MANY of the materials we use everyday are made from rocks and minerals. Pottery mugs, tin cans and glass windows are just three examples. Many paints, especially those used by artists, are made from colorful minerals. You can make your own paints by crushing rocks in a mortar and pestle.

Gold is one of the few metals that is found in its pure state in nature. It is sometimes found as nuggets in rivers. The nuggets can be separated from mud and gravel by panning, and you can try this in the first project below. In the second project, you can experiment with one of the most widely used mineral materials—concrete.

Make paint
Using a mortar and pestle, crush charcoal, brown clay or chalk. Add oil to the powder to make paint.

You will need: gloves, trowel, soil, old wok, water, measuring cup, nuts and bolts preferably made of brass, dishpan.

Gold nuggets
A skilled panner can find single pieces of gold in a whole pan of dirt.

PANNING FOR GOLD

nuggets of "gold"

1 Put on gloves and fill a trowel with soil. Place the soil in an old wok, or shallow pan, along with about a quart of water and some small brass nuts and bolts. Combine it thoroughly using the trowel.

2 Swirl the wok around, letting the soil and water spill over the edge. Add more water if any soil remains and repeat until the water is clear. Panning for real gold washes away the mud, leaving gold behind.

3 When the water is quite clear, examine what is left. You should see that the nuts and bolts have been left behind, as real nuggets of gold would have been, because they are heavier.

PROJECT

MIXING YOUR OWN CONCRETE

1. Place one cupful of sand in a bucket. Add two cups of cement and a handful of gravel. Don't touch the cement with your bare hands.

2. Add water to the mixture little by little, stirring constantly until the mixture has the consistency of oatmeal. Mix it well with the stick.

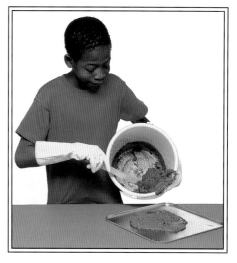

3. Pour the wet concrete onto a tray and spread it out. Let solidify for about half an hour. Wash all the other equipment immediately.

You will need: gloves, measuring cup, sand, bucket, cement, gravel, water, stirring stick, tray, small thin box, foil.

hand prints

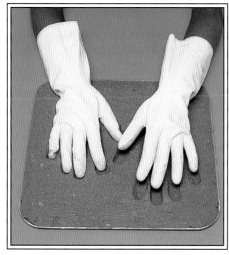

4. At this stage you can shape the concrete any way you want to. Make impressions of your hands, or write with a stick. The marks will be permanent once the concrete has set. Do not use your bare hands.

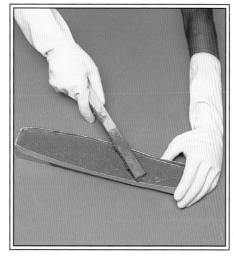

5. You could make a solid concrete block like those used in the construction industry. First, line a small but strong box with foil. Pour in the concrete and smooth the top with a stick.

6. How strong is the concrete block once it has set? Test its strength by trying to bend it. Can you rest a heavy weight on the block?

PLANET EARTH

THE UNIQUE PLANET

THE EARTH is the third planet out from the Sun, the third of the nine planets that make up the solar system. It is a big round ball of rock with a metal core, wrapped in a thin blanket of colorless gases called the atmosphere. From a distance, Earth shimmers like a blue jewel in the darkness of space, because more than 70 percent of its surface is covered with water. No other planet has this much water on its surface. Jupiter's moon Europa is the only other place with much water, but it is so far away from the Sun that the water is frozen. Earth is neither too near the Sun that water is turned to steam, nor too far that it freezes. Only at the poles, is water permanently frozen in ice caps. In places, rock sticks up above the ocean waters to form half a dozen large continents and thousands of smaller islands.

Round Earth
In the last 40 years, spacecrafts have been able to take photographs of Earth from space, so we can see that it is round. The Ancient Greeks suspected that the Earth was round 2,500 years ago, because they saw how ships gradually disappeared over the horizon. But for 2,000 years, many people continued to believe that it was flat. Only when the ship of explorer Ferdinand Magellan sailed around the Earth in 1522 were people finally convinced.

Water world
An island in the Indian Ocean demonstrates all that is unique about planet Earth. The combination of vegetable and plant life, land and sea is found only on Earth—it exists nowhere else in the known universe. Life on Earth exists because it is a watery planet. Water is still, however, a precious resource. Almost 97 percent is salt water in the oceans, and three-quarters of the remaining fresh water is frozen.

The solar system
Earth's nearest companions in space are the planets Venus and Mars, all of which are much the same size. Tiny Mercury nearest the Sun, and even tinier Pluto farthest away, are the terrestrial (rocky) planets of the solar system. The other four planets—Jupiter, Saturn, Uranus and Neptune—are gigantic by comparison and are made not of rock, but gas. It was once thought the solar system was unique in the universe, but astronomers have spotted planets circling distant stars. So there may be another Earth out there after all.

Mercury

Venus

Earth

Mars

The Sun

FACT BOX

- The Earth is not quite round. It bulges out at the equator, making it slightly tangerine-like in shape.

- The Earth's diameter is 7,929 mi. around the equator but only 7,000 mi. around the North and South Poles.

- The Earth takes just a year to travel over 583 billion mi. around the Sun, averaging 66,500 mph.

Man on the Moon

On July 20, 1969, the American astronaut Neil Armstrong stepped from the landing module of the *Apollo 11* spacecraft onto the Moon's surface. It was the first time that a human had ever set foot on the moon. It was an exciting experience, and it reminded the astronaut just how special the Earth is. The Moon is really very close to us in space, yet it is completely lifeless, a desert of rock with no atmosphere to protect it from the Sun's dangerous rays and no water to sustain life.

The surface of Mars

Mars is similar to Earth in size, so people hoped it might have life of its own. Astronomers were once convinced that marks on the surface were canals built by Martians. Sadly, space probes to Mars have found no verifiable signs of life. As this photo from the *Pathfinder* mission shows, it is nothing but rocks and dust. But a few years ago, NASA scientists found what could be a fossil of a microscopic organism in a rock that fell from Mars.

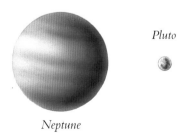

PROJECT

SPINNING PLANET

You will need: *felt-tipped pen, plastic ball, piece of thin string, modeling clay, flashlight.*

NIGHT AND DAY

THE EARTH is like a giant ball spinning in the darkness of space. The only light falling on it is the light of the Sun glowing 93 million miles away. As it spins, the Earth also moves around the Sun. The two ways of moving explain why night and day, and the seasons, occur. At any one time half the world is facing the Sun and is brightly lit, while the other half is facing away and is in darkness. As the Earth spins on its axis, the dark and sunlit halves move around, bringing night and day to different parts of the world.

The Earth is always tilted in the same direction. So when the Earth is on one side of the Sun, the Northern Hemisphere (the area north of the equator) is tilted towards the Sun, bringing summer. At this time, the Southern Hemisphere (the area south of the equator) is tilted away, bringing winter. When the Earth is on the other side of the Sun, the Northern Hemisphere is tilted away, bringing winter, while the south is in summertime. In between, as the Earth moves around to the other side of the Sun, neither hemisphere is tilted more than the other one toward the Sun. This is when spring and autumn occur. These two experiments show how this happens. The ball represents the Earth and the flashlight is the Sun.

1 Draw, or cut out and glue, a shape on the ball to represent the country you live in. Stick the string to the ball with modeling clay. Tie the string to a rail or bar, such as a towel bar, so that the ball hangs freely.

2 Shine the flashlight on the ball. If your country is on the shadow half of the ball on the far side, then it is night because it is facing away from the Sun.

3 Your home country may be on the half of the ball lit by the flashlight instead. If so, it must be daytime because it is facing the Sun. Keep the flashlight level, aimed at the middle.

4 Turn the ball from left to right. As you turn the ball, the light and dark halves move around. You can see how the Sun comes up and goes down as the Earth turns.

PROJECT

THE SEASONS

You will need: *felt-tipped pen, plastic ball, bowl just big enough for the ball to sit on, flashlight, books or a box to set the flashlight on.*

1 Use the felt-tipped pen to draw a line around the middle of the ball. This represents the equator. Sit the ball on top of the bowl so that the equator line is sloping gently.

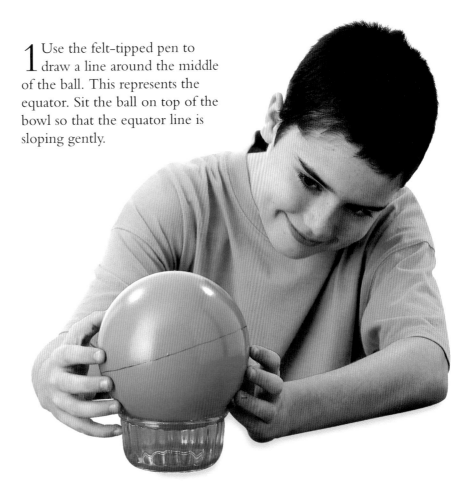

2 Put the flashlight on the books so it shines just above the equator. It is summertime on the half of the ball above the equator where the flashlight is shining, and winter in the other half of the world.

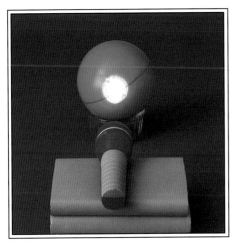

3 Shine the flashlight on the equator, as shown. It sheds an equal amount of light in each hemisphere. This is the equivalent of spring and autumn, when days and nights are of similar length throughout the world.

Half Moon

The Moon shines not because it gives out light itself but because it reflects the light of the Sun. Just as half of the Earth is always lit by the Sun, so is the other half by the Moon. The Moon appears to change shape during the month, from a crescent to a disk and back again. This is because it moves around the Earth and so we see its sunlit side from different angles. When we see a full Moon, we are seeing all of the sunlit side. The Sun and stars are our only sources of light from space.

EARTH STORY

THE SOLAR system formed 4,570 million years ago from debris left over from the explosion of a giant star. As the star debris spun around the newly formed Sun, it began to congeal into balls of dust. Quickly, the dust clumped into tiny balls of rock called planetesimals, and the planetesimals clumped together to form planets such as the Earth and Mars. At first, the Earth was little more than a ball of molten rock. Then, when the Earth was about 50 million years old, it is thought that a giant rock cannoned into it with such force that the rock melted. The melted rock cooled to become the Moon. The Earth itself was changed forever. The shock of the impact made iron and nickel collapse to Earth's center, forming a core so dense that atoms fuse in nuclear reactions. These reactions have kept the Earth's center ferociously hot ever since. The molten rock formed a thick mantle around the core, kept slowly churning by the heat.

A volcanic landscape
It is hard to imagine what the Earth's surface was like in the very early days, but you could get a good idea by staring down into the mouth of an active volcano. The collision of rocks that actually created the Earth left it incredibly hot—hot enough to melt the rock it was made from. The whole planet was just one giant, seething red hot ball of magma (molten rock).

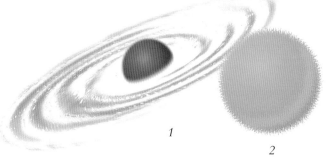
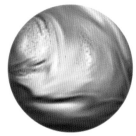

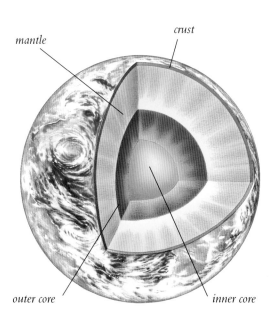

The formation of planet Earth
The solar system and planet Earth formed when gravity began to pull the debris left over from a giant star explosion into clumps (1). At first, Earth was just a molten ball, and for half a billion years it was bombarded by meteorites (2, 3). By 3.8 billion years ago, things were calming down (4, 5). The crust and atmosphere had formed, and very solid lumps of rock on the planet's surface were forming the first continents (6).

The layers of the Earth
The Earth is not a solid ball of rock. The collision with the giant rock that formed the Moon caused the Earth to separate into several distinct layers. On the outside is a thin layer of solid rock, up to 25 mi. thick, called the crust. Below the crust is a thick layer of soft, semi-molten rock known as the mantle. At the center is a core of iron and nickel. The outer core is so hot it is molten, but the pressure in the inner core is so intense it cannot melt, even though temperatures here reach about 6,700°F.

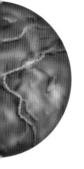

Steam power
These hot springs in Iceland are sending out jets of steam heated by the Earth's interior. But they are tiny compared to the huge amounts of steam and gas that must have billowed out from the hot Earth in its early days. Two hundred million years after the birth of planet Earth, volcanic fumes formed the atmosphere. All that was missing was oxygen—the vital ingredient added later by plants.

Auroras
The spinning of the Earth swirls the iron in its core, turning it into a huge electric dynamo—and the electricity turns the Earth into a magnet. The effect of Earth's magnetism is felt tens of thousands of miles out in space. Indeed, there is a giant cocoon of magnetism around the Earth called the magnetosphere. This shields us from harmful electrically charged particles streaming from the Sun. There are small gaps in this shield above the poles. Here, charged particles collide with the atmosphere and light up the sky in spectacular displays of light called auroras.

PROJECT

MAGNETIC EARTH

The sailor's guide
Until the age of satellites, the magnetic compass was the sailor's main tool for finding their way at sea—working in any weather, at any time of day and on a rolling ship.

MAKE A COMPASS
You will need: bar magnet, steel needle, slice of cork, tape, small bowl, pitcher of water.

THE EARTH behaves as if there is a giant iron bar magnet running through its middle from pole to pole. This affects every magnetic material that comes within reach. If you hold a magnet so that it can rotate freely, it always ends up pointing the same way, with one end pointing to the Earth's North Pole and the other to the South Pole. This is how a compass works—the needle automatically swings to the north. These projects show you how to make a compass, and how you can use it to plot the Earth's magnetic field. The Earth's magnetic field is slightly tilted, so compasses do not swing to the Earth's true North Pole, but to a point that is a little way off northern Canada. This direction is known as magnetic north.

The Earth's magnetism comes from its inner core of iron and nickel. Because the outer core of the Earth is liquid and the inner core solid, the two layers rotate at different rates. This sets up circulating currents and turns it into a giant dynamo of electrical energy.

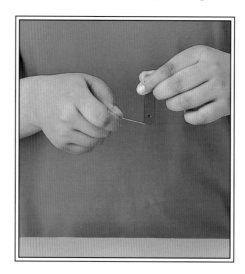

1 To turn the needle into a magnet, stroke the end of the magnet slowly along it. Repeat this in the same direction for about 45 seconds. This magnetizes the needle.

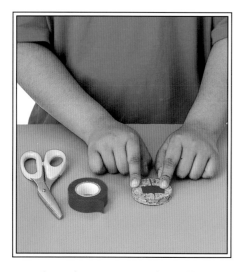

2 Place the magnetized needle on the piece of cork. Make sure that it is exactly in the middle, otherwise it will not spin evenly. Tape the needle into position.

3 Fill the bowl nearly to the top with water and float the cork in it. Make sure the cork is exactly in the middle and turns without rubbing on the edges of the bowl.

4 The Earth's magnetic field should now swivel the needle on the cork. One end of the needle will always point to the north. That end is its north pole.

PROJECT

MAGNETIC FIELD

You will need: *large sheet of paper, bar magnet, your needle compass from the first project, pencil.*

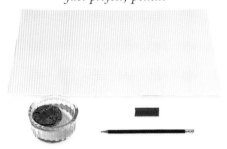

1. Lay a large sheet of paper on a table. Put the magnet in the middle of the paper. Set up your needle compass an inch away from one end of the magnet.

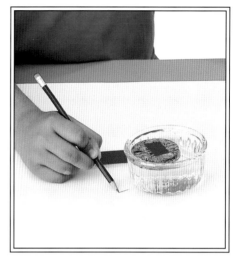

2. Wait as the compass needle settles in a particular direction as it is swiveled by the magnet. Make a pencil mark on the paper to show which way it is pointing.

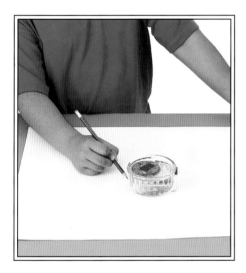

3. Move the compass a little way toward the other end of the magnet. Mark a line on the paper to show which way the needle is pointing now.

4. Repeat Step 3 for about 25 different positions around the magnet. Try the compass both near the magnet and further away. You should now have a pattern of marks.

Magnetic protection

A vast region of space about 38,000 mi. out from Earth is full of electrically charged particles. This area is called the magnetosphere. Without its protection, Earth would be exposed to the solar wind, a lethal stream of charged particles moving rapidly from the Sun.

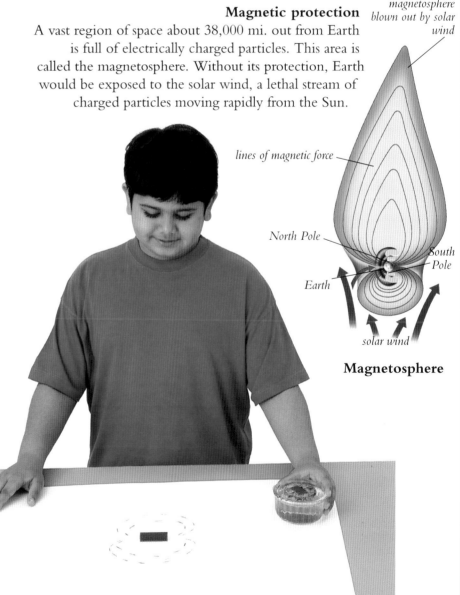

tail of magnetosphere blown out by solar wind

lines of magnetic force

North Pole

South Pole

Earth

solar wind

Magnetosphere

73

LIFE-SUPPORT SYSTEM

LIFE ON Earth exists only because of the atmosphere. This thin blanket of gases, tiny water drops and dust is barely thicker on the Earth than the skin is on an orange. At its thickest, it is less than 600 miles deep. The atmosphere has no color or taste, yet it is much more interesting than it may appear, because it is a surprisingly complex mixture of gases. Over 99 percent of the atmosphere is made up of two gases, nitrogen (78 percent) and oxygen (21 percent). It also contains argon, carbon dioxide, water vapor and minute traces of many other gases, such as helium and ozone.

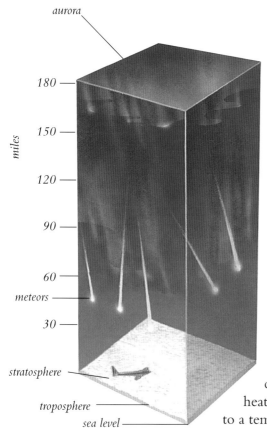

Layers of air
We live in the densest, bottom layer of the atmosphere, called the troposphere. This is where most of the water vapor is, and all the weather. The calmer stratosphere is 7 to 30 mi. above ground, and gets warmer higher up. The mesosphere above is thinner in gases and very cold. Beyond is the thermosphere, heated by ultraviolet rays from the Sun to a temperature of 3,600°F.

Without the atmosphere, planet Earth would be as lifeless as the Moon. The atmosphere gives us air to breathe and water to drink, and keeps us warm. The atmosphere also acts as a shield, filtering out the Sun's harmful rays and protecting us from falling meteorites.

FACT BOX

• If all the water vapor in the air suddenly condensed, it would make the oceans 1 in. deeper.

• Over a billion tons of salt join the atmosphere from sea spray each year, and over a quarter of a billion tons of soil dust.

• The stratosphere glows faintly at night. This airglow is created by sodium or salt spray from the sea that is heated by the Sun during the day.

• Early radio signals were bounced around the world off electrically charged particles in the thermosphere.

A thin veil
Earth's atmosphere can be seen in this photograph taken from the space shuttle. The Sun is beyond the horizon, just catching the edge of the new moon high above. As the Sun's rays are scattered around, through the dense, watery, dusty lower layers of the atmosphere, they turn red and orange. Further up, where the air is thinner and there is no dust or water, the atmosphere is blue. This is what you see when you look up on a clear day. The sky is blue because gases in the air reflect mostly blue light from the Sun. Dust and moisture scatter other colors and dilute the blue.

Stratocruiser

Airplanes climb high up above 7 mi. as soon as they take off. By doing this, they break out of the troposphere, where they would be buffeted by the weather, and soar into the calm stratosphere, where there is no weather at all. At this height, the passengers need to be inside a pressurized cabin. Our bodies are built to cope with the pressure of air at ground level. Up in the stratosphere where the air is thin, the pressure is too low.

Outer limits

The space shuttle orbits Earth in space about 180 mi. up. This is the outermost layer of Earth's atmosphere, the exosphere. Here the gases are rarefied; that is, they are few and far between. Above the exosphere, the atmosphere fades away into empty space. There is no oxygen at this height—just nitrogen and more of the lighter gases such as hydrogen.

Thin air

Gravity pulls most of the gases in the air into the lowest layers of the atmosphere. Seventy-five percent of the weight of the gases in the air is squashed into the troposphere, the lowest one percent of the atmosphere. The air gets thinner very quickly as you climb. Mountaineers climbing the world's highest peaks need oxygen masks to breathe because there is much less oxygen at this height.

Making clouds

Clouds are made from billions of tiny droplets of water and ice crystals so tiny that they float on the air. The droplets are formed from water vapor rising from the sea as it is warmed by the Sun. Air gets steadily colder higher up, and as the vapor rises it cools down. Eventually the water vapor gets so cold that it turns into tiny droplets of water and forms a cloud.

STIRRINGS OF LIFE

LIFE PROBABLY began on Earth entirely by chance, about 3.8 billion years ago. The early Earth was a hostile place. It seethed with erupting volcanoes, was washed by oceans of warm acid and enveloped in toxic fumes. But these could have been just the right conditions to start life. Amino acids, the building blocks from which the first living cells were formed, may have been created by chance as small molecules were fused together by lightning bolts that surged through the stormy air.

Recently, amazing microscopic bacteria called archaebacteria were found living on black smokers on the ocean floor, in conditions as hostile as the early Earth. It may be that bacteria such as these were the first living cells, feeding on the chemicals spewed out by volcanoes. Another kind of bacteria, called cyanobacteria, or blue-green-algae, appeared later.

Geological time

Much of what we know about the story of life on Earth comes from fossils, which are the remains of organisms in rocks. Layers of rock form one on top of each other, so the oldest is usually at the bottom, unless they have been disturbed. By studying rocks from each era, paleontologists (scientists who study fossils) have slowly built up a picture of how life has developed over millions of years.

Ancient life-form

Archaebacteria are the oldest known form of life. They seem to be able to thrive in the kind of extreme conditions that would kill almost everything else, including other bacteria. Some have been found living in boiling sulfur fumes on volcanic vents on the sea floor. This one was found in Ace Lake, Antarctica, and can survive in incredibly cold temperatures, living off carbon dioxide and hydrogen.

Hardy organisms

One of the most remarkable discoveries of recent years has been tall volcanic chimneys on the ocean floor that belch hot black smoke. These black smokers, or hydrothermal vents, are home to a community of amazing organisms that actually thrive in the scalding waters and toxic chemicals that would kill other creatures. These communities give important clues to how life could have started in the similarly hostile conditions on the primeval Earth four billion years ago.

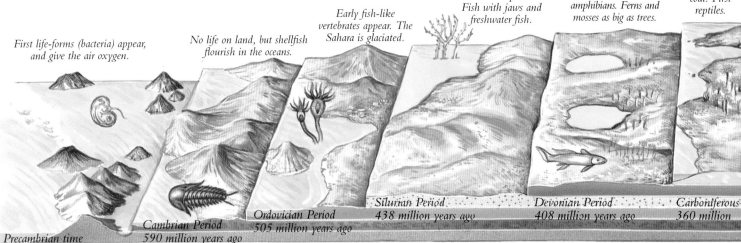

First life-forms (bacteria) appear, and give the air oxygen.

No life on land, but shellfish flourish in the oceans.

Early fish-like vertebrates appear. The Sahara is glaciated.

First land plants. Fish with jaws and freshwater fish.

First insects and amphibians. Ferns and mosses as big as trees.

Vast warm swamps of fern forests which form coal. First reptiles.

Precambrian time

Cambrian Period 590 million years ago

Ordovician Period 505 million years ago

Silurian Period 438 million years ago

Devonian Period 408 million years ago

Carboniferous 360 million

Puffs of oxygen

Early archaebacteria left no trace. The oldest signs of life are microscopic threads in rocks dating back 3.5 billion years. The threads are like the blue-green algae called cyanobacteria that live in the oceans today. The tiny algae changed the world by using sunlight to break carbon dioxide in the air into carbon and oxygen. They fed on the carbon and expelled oxygen. The little puffs of oxygen seeped into the air, filling it with oxygen and preparing the way for life as we know it.

Ancient slime

The oldest proofs of life are stony mounds called stromatolites. Some of those at Fig Tree Rock in South Africa, date back 3.5 billion years. Stromatolites are the fossilized remains of huge colonies of slimy bacteria with a thin layer of cyanobacteria on top. The cyanobacteria obtain their food from sunlight, while the bacteria feed on dead cyanobacteria. Stromatolites called conyphytons grew up to 300 ft high.

Clouds of life

Most scientists believe that the chemicals of life were assembled on Earth. Some, such as the late Fred Hoyle, believe life came from space. Clouds of stardust in space, called giant molecular clouds, do contain basic life chemicals. These include huge amounts of ethyl alcohol—the alcohol in drinks.

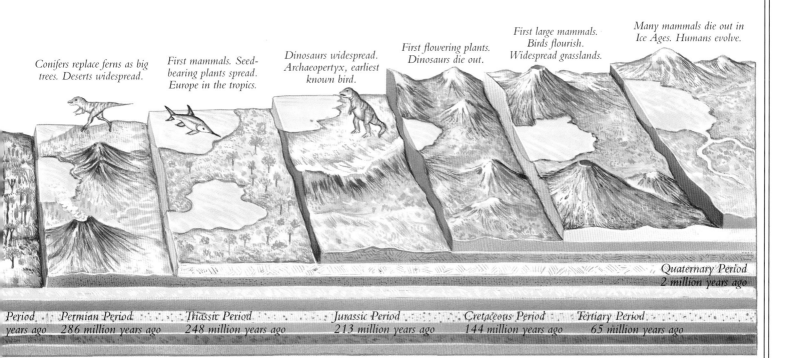

Conifers replace ferns as big trees. Deserts widespread.

First mammals. Seed-bearing plants spread. Europe in the tropics.

Dinosaurs widespread. Archaeopertyx, earliest known bird.

First flowering plants. Dinosaurs die out.

First large mammals. Birds flourish. Widespread grasslands.

Many mammals die out in Ice Ages. Humans evolve.

Permian Period — 286 million years ago
Triassic Period — 248 million years ago
Jurassic Period — 213 million years ago
Cretaceous Period — 144 million years ago
Tertiary Period — 65 million years ago
Quaternary Period — 2 million years ago

EVOLUTION OF SPECIES

Ammonites
Most fossils that are found are shellfish, because they are so easily preserved. One of the most common shellfish fossils is the ammonite, a creature a bit like a squid that lived inside a spiral shell. Ammonites lived in the oceans around the time when dinosaurs dominated the land—from 220 to 65 million years ago. They are very common and evolved quickly into different types, that can be slotted into specific periods of time. This means that geologists can use them to date the rocks they are found in.

For the first three billion years of Earth's history, the only life was in the form of microscopic, single-celled organisms in the oceans. Then, about 700 million years ago, the first real animals appeared. These were creatures such as jellyfish and sponges made from many kinds of cells, each one suited to a different task. These creatures were soft and left few traces. Over the next 100 million years, animals with shells and bones appeared. Hard parts fossilize easily, and from this time—the start of the Cambrian Period about 590 million years ago—there are very many fossils. From these, scientists have pieced together the story of how different species have come and gone. These include not only the dinosaurs, the gigantic reptiles that dominated the planet for 155 million years, but the first human-like creatures.

First plants
Life began in the oceans and moved onto land only gradually from about 400 million years ago. The first trees were not like today's trees. They were the ancestors of today's ferns and cycads, like this one in Australia. Unlike modern trees, which grow from seeds and flowers, ferns and cycads grow from tiny spores. Today, ferns and cycads are usually quite small, but about 300 million years ago, they grew into huge, tall forests.

Living fossils
Coelacanths are remarkable fish that first appeared 400 million years ago. They were once thought to have died out 65 million years ago, since no more recent fossils have been found. But in 1938 a living coelacanth was found in the Indian Ocean. The fish have muscular, limb-like fins, and it is from species like these that the first land creatures evolved. The limb-like fins developed into legs so that the fish could haul themselves across mud flats.

FACT BOX

• The first creatures with bones appeared 570 million years ago.

• The first plants and insects appeared on land 400 million years ago.

• The first amphibians crawled onto land 350 million years ago.

• The biggest carnivore, the sea reptile lipluridon, lived 150 million years ago and grew to 80 ft long.

• The biggest land creature, the dinosaur Brachiosaurus, grew to 80 ft long.

• Evolution was not a smooth, continuous process but was broken by periodic mass extinctions of species.

Lifestyle

Dinosaurs dominated Earth from 225 to 65 million years ago, and then they mysteriously died out altogether. As paleontologists have found more and more dinosaur fossils, they have been able to figure out in detail what they looked like and how they lived. Stegosaurs were plant-eating dinosaurs that lived from about 180 to 80 million years ago. They had two rows of pointed plates down their back and a spiny tail that they swung to protect themselves from predators.

Frozen baby

Woolly mammoths were hairy, elephant-like creatures that lived in northern Asia and North America until about 10,000 years ago. Mammoths were probably hunted to extinction by early humans, but bodies have been found frozen in the permafrost (permanently frozen ground) of Siberia. They are so well preserved that some Japanese scientists hope to recreate them. The idea is to take from the bodies DNA, the chemical that carries their genetic code. The DNA could be used to create an embryo, which could be implanted in a living elephant.

Digging for life

When miners dig coal out of a mine, they are digging out the history of life on Earth. Coal is the fossilized remains of forests of giant club mosses and tree ferns that grew in vast tropical swamps some 350 million years ago, in the Carboniferous Period. Over millions of years, layer upon layer of dead plants sank into the swamp mud. As they were buried deeper and deeper, they were squeezed dry and became hard, slowly turning to almost pure carbon. The deeper the remains were buried, the more completely they have turned to carbon. So, coal near the surface is brown. Deep down it is jet black.

PROJECT

COMPETITIVE GENES

You will need: *medium-size bowl, spoon, 3½ oz sea salt, liquid fish food, brine shrimp eggs from a pet store, magnifying glass.*

TODAY, THERE is an astonishing variety of life on Earth, with millions of species of animals and plants. Yet each one has its own natural home or way of life. Every living thing is adapted (suited) to its surroundings. In 1859, the English naturalist Charles Darwin explained all this with his theory of evolution by natural selection. This theory shows how over millions of years species gradually change or evolve. As they change, they adapt to their surroundings, and new species emerge. Evolution like this depends on the fact that no two living things are alike. So some may start life with features that make them better able to survive, as the first experiment shows. An animal, say, might have long legs to help it escape from predators. Individuals with such valuable features have a better chance of surviving. They may also have offspring that inherit these features, as in the way shown in the Strong Genes experiment. Slowly, over generations, better adapted animals and plants flourish, while others die out or find a new home. In this way, species gradually evolve.

Non-survivor
The ancient stingray fossilized here survived for millions of years. Then, this ancient species suddenly became extinct. Conditions, such as the climate, changed, and the fish did not adapt to the new conditions fast enough to survive.

COMPETING FOR LIFE

1 To see how some eggs survive and others don't, pour ⅓ gallon of warm water into a bowl. Stir in the sea salt until it dissolves. When the water is cool, add a few drops of fish food.

2 Sprinkle in a spoonful of shrimp eggs. Set the bowl in a warm place so that the water will stay at 70°F.

3 After a few days, some shrimps will hatch into larvae. Stir the water once a day and scoop out a spoonful. Be careful not to disturb the larvae too much.

4 Using a magnifying glass, count how many eggs and larvae you see on the spoon. When adults appear, count these too. Only the strongest will survive and grow into adults.

PROJECT

The lottery of life
Only a tiny proportion of this dragonfly's thousands of eggs will live to adulthood. Dragonflies generally have adapted well to changing conditions. They are the longest-surviving of all insect species. More than 300 million years ago, they were the first animals to fly.

STRONG GENES
You will need: pack of ordinary playing cards. This is a game for three or more players.

1 Deal seven cards to each player. The cards are your supply of genes for life. The suit of diamonds represents strong genes that give you a better a chance of survival.

2 Each player lays down a card in turn, following suit. If you cannot follow suit, play a diamond—a strong gene. If you play a diamond, always pick it up and save for the next deal.

3 The player who plays the highest card (or a diamond) wins the round, and begins the next. Once all the cards have been played, the player who has won the fewest rounds drops out.

4 Deal six cards to each survivor. Each player then picks up their diamonds from the previous round. Discard one card from the six for each diamond you pick up.

5 Play out all the cards in rounds as before. Again the player who wins the fewest rounds must drop out. Repeat Step 4 dealing five cards each. Play again with four cards, then three, then two, then one. The last player survives life's game.

BROKEN EARTH

THE SURFACE of the Earth is not in one piece, but cracked like a broken eggshell, into 20 or so giant slabs. The giant slabs, called tectonic plates, are huge, thin pieces of rock thousands of miles across but often little more than a dozen miles thick. The plates are not set in one place, but are slipping and sliding around the Earth all the time. They move very slowly—at about the pace of a fingernail growing—but they are so gigantic their movement has dramatic effects on the Earth's surface. The movement of the plates causes earthquakes, pushes up volcanoes and mountains and makes the continents move. Once, the continents were all joined together in one huge continent that geologists call Pangea, surrounded by a giant ocean called Panthalassa. About 200 million years ago the plates beneath Pangea began to split up and move apart, carrying fragments of the continent with them. These fragments slowly drifted to the positions they are in now.

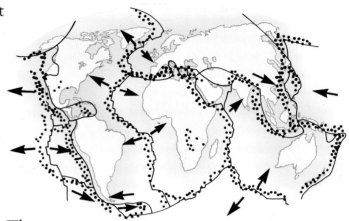

Surface sections
The Earth's rigid shell is called the lithosphere, from the Greek word *lithos* (stone). It is broken into the huge fragments shown on this map. The African plate is gigantic, underlying not only Africa but half of the Atlantic Ocean too. The Cocos plate under the West Indies is quite small. Black dots mark the origins of major earthquakes over a year. Note how they coincide with the plate margins.

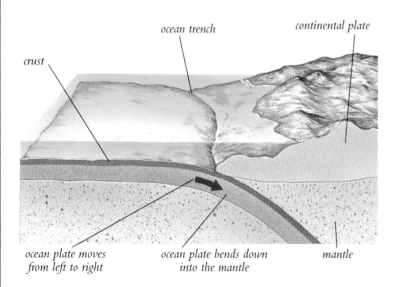

Plates in collision
In many places, tectonic plates are slowly crunching together with enormous force. As they collide, one plate may be forced under the other in a process called subduction, which means drawing under. The plate is completely destroyed as the plate slides down into the heat of the Earth's mantle. Earthquakes are often generated as it slides down, and the melting rock bubbles up as violent volcanoes. Subduction zones occur right around the western edge of the Pacific Ocean.

Lowest place on Earth
When one plate is forced down beneath another, it can open up deep trenches in the ocean floor. These trenches are the lowest places on the Earth's surface. They plunge so far down that the water is darker than the blackest night. These dark abysses have been never been fully explored, but this photograph was taken at the bottom of the world's deepest trench, the Marianas Trench, at a depth of over 32,000 ft.

Two-way friction

In some places, the plates are neither crunching together nor pulling apart. Instead, they are shaking sideways past each other. This is happening at the San Andreas fault in California. The giant Pacific plate is sliding at a rate of 2½ in. a year northwest past the North American plate. This has set off earthquakes that have rocked the cities of San Francisco and Los Angeles.

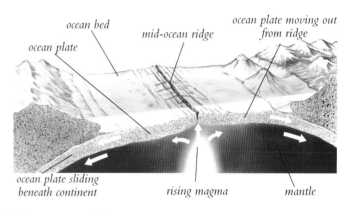

Mid-ocean ridge

Right down the middle of the seabed in the Atlantic Ocean, there is a giant crack where the tectonic plates are pulling apart. Here, molten rock from the Earth's interior wells up into the crack and freezes onto the edges of the plates. This creates a series of jagged ridges called the mid-ocean ridge. As the plates move apart and new rock is added on the edges, so the sea floor spreads wider.

A string of islands

As one colliding plate is forced down beneath another into the Earth's hot interior, it melts. The melting rock is squeezed into cracks in the overlying plate and bursts onto the surface as a line of volcanoes. If this line is in the sea, the volcanoes form a long, curving string of islands called an island arc. Many of the islands in the western Pacific formed in this way.

FACT BOX

- The west coast of Africa looks as though it would slot into the east coast of South America because they were once joined together.

- New York is moving ¾–1¼ in. away from London every year.

BUILDING MOUNTAINS

Lone volcano
Mount Kilimanjaro, on the border of Kenya and Tanzania, is Africa's highest mountain, reaching 19,340 ft. This distinctive volcanic cone rises in isolated majesty from the surrounding grassland. It was formed where a rising plume of hot rock in the Earth's interior burned its way up through the tectonic plate that underlies East Africa. Mount Kilimanjaro is high enough to be snowcapped, even though it lies near the equator.

A FEW of the world's highest mountains, such as Mount Kilimanjaro in Africa, are lone volcanoes, built up by successive eruptions. Most high mountains are part of great ranges that stretch for hundreds of miles. Mountains look as though they have been there forever, but geologically they are quite young. They have all been created in the last couple of hundred million years—the last quarter of the world's history—by the huge power of the Earth's crust as it moves. The biggest ranges, such as the Himalayas and the Andes, are fold mountains. These are great piles of crumpled rock pushed up by the collision of two of the great plates that make up the Earth's surface. Folding opens up many cracks in the rock and the weather attacks them, etching the mountains into jagged peaks and knife-edge ridges. Some mountain ranges, in the central parts of plates, are huge blocks of the Earth's crust that have been pushed up as the plate was stretched.

Bow-wave
Geologists used to think that rock crumples to form fold mountains, such as the Alps, in rigid layers. Now they are beginning to think that because this all happens so slowly, the rock flows almost like very thick molasses. They suggest that as one tectonic plate collides with another, the rock is pushed up like the v-shaped bow-wave in front of a boat. Like a very slow bow-wave, the mountains are continually piling up in front of the plate and flowing away at the side.

Continental crunch
The incredible contortions created by rock folding even on a small scale are clearly visible in the zigzag layers of shale. These rocks in Devon, England were folded about 250 million years ago. This was when a mighty collision between two continents that formed the single giant continent of Pangea. Present-day Devon was one of the places that was squeezed in the gap.

Waves in the Andes
Using the latest satellite techniques, geologists have surveyed many of the world's high mountain ranges, including the Andes, the Himalayas, the European Alps and New Zealand. What they have found is something remarkable. When compared to surveys from a century ago, mountain peaks in these ranges have moved exactly as though they are flowing very slowly. So the folds in the rock you see in this photograph may not be folds but very, very slow and stiff waves.

How faults occur

The slow, unstoppable movement of tectonic plates puts rock under such huge stress that it sometimes cracks. Such cracks are called faults. Where they occur, huge blocks of rock slip up and down past each other, creating cliffs. In places a whole series of giant blocks may be thrown up together, creating a new mountain range. The Black Mountains in Germany are an example of block mountains formed in this way.

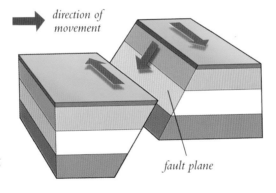

direction of movement

fault plane

The Earth splits open

The Great Rift Valley runs for more than 1,800 mi. down the east of Africa from the Red Sea. The movement of the Earth's tectonic plates not only builds up mountains, it also opens up great valleys. Where plates are pulling apart, or magma is pushing up underneath, the land can split like the crust of a pie in the oven. Land can drop along this crack to form a rift valley.

Zagros Mountains, Iran

For the last 100 million years, the tectonic plates carrying Africa and Arabia have been plowing northward into Eurasia. The tremendous impact has crumpled the edge of Eurasia and thrown up a belt of mountains stretching all the way from southern Europe, through the Aegean and Turkey to the Zagros Mountains of Iran. The Turkish and Iranian end remains very active, building up the mountains still higher, and setting off earthquakes.

PROJECT

FOLDING TECHNIQUES

ROCKS TEND to form in flat layers called strata. Some, called sedimentary rocks, form when sand and gravel and seashells settle on the seabed. Volcanic rocks form as hot molten rock streams from volcanoes that flood across the landscape.

Although rock layers may be flat to begin with, they do not always stay that way. Most of the great mountain ranges of the world began as flat layers of rock that crumpled. They were crumpled by the slow but immensely powerful movement of the great tectonic plates that make up the Earth's surface. Mountains occur mostly where the plates are crunching together, pushing up the rock layers along their edges into massive folds. These two experiments help you to understand just how that happens. Sometimes, the folds can be tiny wrinkles just an inch or so long in the surface of a rock. Sometimes, they are gigantic, with hundreds of miles between the crests of each fold. As the layers of rock are squeezed horizontally, they become more and more folded. Some folds turn right over to form overlapping folds called nappes. As nappes fold upon each other, the crumpled layers of rock are raised progressively higher to form mountains. Sometimes you can see the complicated twists of folds on the side of a mountain or cliff.

Mountain range
Although many mountain ranges are getting higher at this very moment as plates move together, mountain building is thought to have been especially active at certain times in Earth's history known as orogenic phases, each lasting millions of years.

SIMPLE FOLDS
You will need: thin rug.

1 Find an uncarpeted floor and lay the rug with the short, straight edge up against a wall. Make sure the long edge of the rug is at a right angle to the wall.

2 Now push the outer edge of the rug toward the wall. See how the rug crumples. This is how rock layers buckle to form mountains as tectonic plates push against each other.

3 Push the rug up against the wall even more and you will see some of the folds turn right over on top of each other. These are folded-over strata or layers, called nappes.

PROJECT

COMPLEX FOLDS

You will need: *rolling pin, different colors of modeling clay, modeling tool, two blocks of wood measuring 2-in. square, two bars of wood measuring 4 x 2 in.*

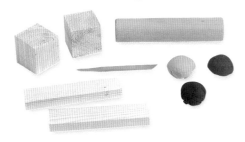

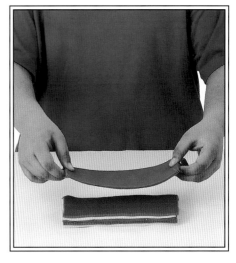

1 Roll out the modeling clay into flat sheets, each about ¼ in. thick. Cut the clay into strips about same width as the blocks of wood. Square off the ends.

2 Lay the plasticine strips carefully one on top of the other, in alternating colors or series of colors. These strips represent the layers of rock.

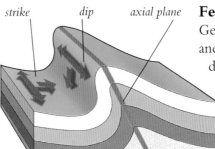

strike   dip   axial plane

Features of a fold

Geologists describe an upfold as an anticline and a downfold as a syncline. The dip is the direction the fold is sloping. The angle of dip is how steep the slope is. The strike is the line along the fold. The axial plane is an imaginary line through the center of the fold—this may be vertical, horizontal or at any angle in between.

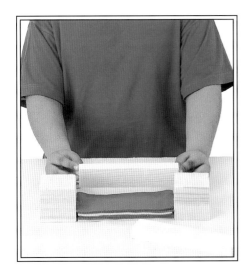

3 Place the blocks of wood at both ends of the strips of plasticine. Lay the bars of wood down both sides of the strips to stop them from twisting sideways.

5 Occasionally, stop and pull away the bars of wood to see what is happening. As you push harder, see how the layers crumple increasingly.

4 Ask a friend to hold on to one block while you push the other towards it. As you push, the effect is similar to two tectonic plates slowly pushing together.

WEARING AWAY

IF YOU went to the Moon, you would see the footprints of the first astronauts to set foot there way back in 1969. They are still there because the Moon's surface never changes. It has no atmosphere and its surface is now completely still. Earth's surface changes all the time. Most of these changes take place over millions of years, far too gradually for us to see. Occasionally, though, the landscape is reshaped dramatically and quickly, such as when a volcano throws up a completely new mountain in minutes, or an avalanche brings down the entire side of a hill in seconds. The Earth's surface is shaped in two ways. First, it is distorted and reformed from below by the gigantic forces of the Earth's interior. Second, it is shaped from above by the weather, running water, waves, moving ice, wind and other agents of erosion.

The mountain's end
In mountainous regions, the attack of the weather on rock can break off huge amounts of debris. Millions of years of shattering, especially by frost, can create vast numbers of angular stones called "scree." As the stones fall off steep slopes, they gather at the foot of the slope in huge piles called scree slopes. Eventually, these stones will be broken down. They will form fine sand and silt and gradually wash away down to the sea where they can begin to form new rock.

Cycle of erosion
A century ago, Harvard Professor W. M. Davis (1850–1935) suggested that landscapes are shaped by "cycles of erosion" going through "youth, maturity and old age." This theory gave an idea of how landscapes evolve, but research has shown that the truth is more complex.

Youth: After an uplift of the land, there is vigorous erosion as fast-flowing streams bite deep into the landscape.

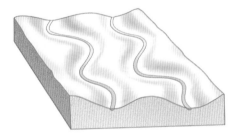

Maturity: River valleys get wider and slopes get gentler as they are worn away. Hill tops are rounded off.

Old age: Valleys are worn flat into wide plains called peneplains and slopes are reduced to isolated hills.

Shattered peaks
As soon as rock is exposed to the weather, it starts to break down under the assault of wind and rain, frost and sunlight. Sometimes the rock is corroded (eaten away) by chemicals in the air, or rainwater trickling over it. Sometimes it is broken down physically by, for example, the effects of heat and cold. Water in cracks can expand so forcefully as it freezes that it can shatter the toughest rock.

Tors and kopjes

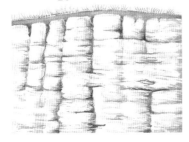
Water trickles down into the ground through joints and cracks in the rock.

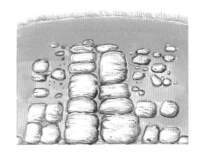
The water corrodes and widens the joints, and the soft debris is washed away.

Eventually, only the big, intact blocks are left perched on the hill top.

In cool regions, there are outcrops called tors on the tops of moors. What is surprising about tors is that they are made from massive blocks that poke above completely smooth slopes, like a castle on a hilltop. There are similar outcrops called kopjes in the tropics. Both features are thought to have been created by the gradual uncovering of rock that has been corroded into big blocks by chemicals in water trickling through the ground, as the pictures above show.

Desert sculpture

In most places, running water is the main agent of erosion. The landscape is molded into rounded hills and deep valleys by the wash of rivers and streams. In deserts, however, running water is sparse, although rivers may flow for a short time after rain and cut valleys. Much of the landscape is sharp and angular and cut into weird shapes by the blast of windblown sand. This is Mexican Hat in the Utah Desert in North America.

FACT BOX

- At -7.6°F, ice can exert a pressure of 6,600 lbs on an area the size of a coin.

- The Colorado River began carving out the Grand Canyon 60 million years ago, as the river plain was slowly uplifted by giant movements of the Earth's crust, forming a plateau.

Acid work

A limestone statue on Gloucester Cathedral in England has worn away. Some rocks are better than others for building materials. All rain is slightly acidic. Limestone rock is especially susceptible to corrosion by acidic rainwater. The rainwater seeps into cracks in the rock, eating it away underground and creating potholes, tunnels and spectacular caverns.

PROJECT

PRACTICAL EROSION

Rocks and mountains look tough, but all the time they are being worn away by the weather, by running water, by waves, by glaciers and by the wind. It is such a slow process that you can rarely see it happening. These three projects speed up the process so that you can see instantly what takes millions of years in nature. The landscape is slowly reshaped, as rocks crumble and mountains are worn down in a process called denudation (laying bare). Much of the damage is done by the weather.

Wherever rock is exposed to the weather, it is attacked by the atmosphere and begins to crumble away. This process is called weathering. Mechanical weathering is when rock is broken down by heat and cold. In areas where the temperature falls below freezing, water seeps into cracks in rocks and then freezes. It expands as it freezes, making the rock shatter or split. Chemical weathering is when the slight acidity of rainwater dissolves rock like tea dissolving sugar. Limestone landscapes can be corroded into fantastic shapes by chemical weathering.

Eroded landscape
Often, the effects of denudation are hidden under a covering of soil and debris. But they are obvious here in this dry landscape where the damage done by the weather and water running off the land is clearly exposed.

THE DESTRUCTIVE POWER OF WATER

You will need: baking sheet, brick, tray or bowl, sand, castle mold, pitcher, water.

1 Put one end of a baking sheet on a brick. Put the other end of the baking sheet on a lower tray or bowl, so that it slopes downward. Make a sandcastle on the baking sheet.

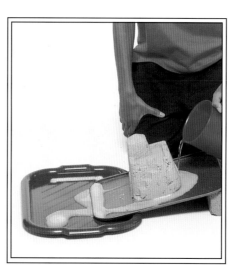

2 Slowly drip water on the castle. Watch the sand crumble and form a new shape. This is because the sand erodes away where the water hits it.

3 Ensure that the water flows down the center of the baking sheet. This way the water hits the middle of the sand castle, eroding the center to form a natural stack.

PROJECT

HARD WATER

You will need: 3 small aluminum foil sheets or saucers, pitcher, mineral water, tap water, distilled water, 3 labels, pen.

1 Fill three foil sheets or saucers with a small amount of water—one with mineral water, one with tap water and one with distilled water.

2 Label the sheets with the type of water in them and set them somewhere warm and well-ventilated. Ensure that they will not be disturbed for a few days.

3 Examine the sheets once the water has evaporated. You will see that the distilled water has not deposited minerals because it does not contain any. Mineral water deposits only a few minerals. Tap water deposits vary depending on where you live. In hard water areas, tap water flows over rocks such as limestone and chalk, and deposits lime minerals. Water in soft water areas flows over rocks such as sandstone. This does not dissolve in water and leave a deposit.

distilled water

tap water

mineral water

CHEMICAL EROSION

You will need: baking sheet, brown sugar, pitcher of water.

1 Build a pile of brown sugar on a sheet. Imagine that it is a mountain made of a soluble rock (dissolves in water). Press the sugar down firmly and shape it to a point.

2 Drip water on your sugar mountain. It will erode as the water dissolves the sugar. The water running off should be brown, because it contains dissolved brown sugar.

RUNNING RIVERS

RIVERS PLAY a large part in shaping the landscape. Without them, the landscapes would be as rough and jagged as the surface of the Moon. Tumbling streams and broad rivers gradually mold and soften contours, wearing away material and depositing it elsewhere. Over millions of years, a river can carve a gorge thousands of yards deep through solid rock, or spread out a vast plain of fine silt. Wherever there is water to sustain them, rivers flow across the landscape. They start high in the hills and wind their way down toward the sea or a lake. At its head, a river is little more than a trickle, a tiny stream tumbling down the mountain slopes. It is formed by rain running off the mountainside or by water bubbling up from a spring. As the river flows downhill, it is joined by more and more tributaries and gradually grows bigger. As it grows, its nature as well as its power change dramatically.

Thundering water
Waterfalls are found where a river plunges straight over a rock ledge and drops vertically. Typically, they occur where the river flows across a band of hard rock. The river wears away the soft material beyond the hard rock and makes a sudden step down. This is Victoria Falls in Zimbabwe, formed where the Zambezi River suddenly drops about 300 ft into a deep, narrow chasm. The spray and the roar of the water have given the Falls the local name *Mosi oa Tunya,* the smoke that thunders.

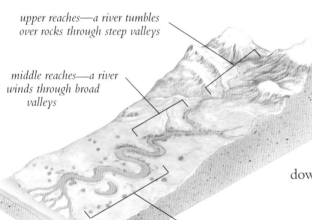

upper reaches—a river tumbles over rocks through steep valleys

middle reaches—a river winds through broad valleys

lower reaches—a river winds broad and smooth across flat floodplains

A river's course
As it flows down to the sea, a river changes its character. In the upper reaches, it is a fast-flowing, tumbling stream that cuts down through steep, narrow valleys. Lower down, a river broadens and deepens. Eventually, it meanders (winds) back and forth across broad floodplains made of material that it has washed down from higher up.

Forceful
High in the mountains, streams are small and tumble down over rocks and boulders. The valley is narrow with steep sides. Boulders often crash down into the stream bed so that the stream is forced to wind its way around them. The flow of water is very erratic. In flat places it flows slowly, whereas in other places it plunges down fast and furiously over rapids and waterfalls. Sometimes, when snow melts, the water level may rise enough to roll big boulders along.

Winding rivers

All rivers wind. As they near the sea, they wind more often, forming horseshoe-shaped bends called meanders, like these in the Guiana Highlands. Meanders begin as the river deposits sediments along its bed in ups and downs called pools (deeps) and riffles (shallows). The distance between pools and riffles, and the size of meanders, is usually in proportion to the width of the river. Meanders develop as the river cuts into the outer bank of a bend and deposits sand and mud on the inner bank.

Colorado loop

Big "gooseneck" bends usually only form when a river is crossing broad plains. But here on the Colorado, a gooseneck is in a deep gorge. A meander that cuts into a gorge in this way is called an incised meander. It probably formed millions of years ago when the Colorado Plateau was a flat lowland. The land was lifted upward and the meander cut deeper as the land rose.

Black river

The Rio Negro, a major tributary of the Amazon in South America, is inky black. This is because of the rotting vegetable matter the river has picked up from the mangrove swamps it flows through. This is why it is called the Negro—*negro* is Spanish for black. Rivers carry their load of sediment in three ways. Big stones are rolled along the river bed. Smaller grains are bounced along the bed. The finest grains float in the water. Typically the load is mostly yellowish silt (fine mud) and sand. The Yangtze River in China carries so much yellow silt that it is often known as the Yellow River.

ICE SCULPTING

GLACIERS ARE "rivers" of slowly moving ice that form in mountain regions where it is too cold for the snow to melt. They flow down mountain valleys, creeping lower and lower until they reach a point where the ice begins to melt.

The ice in a glacier is not clear, but opaque like packed snowballs. The surface is streaked with bands of debris, such as rocks, that has fallen from the mountain slopes above. Massive cracks appear where the ice travels over bumps in the valley floor. Today, glaciers form only in high mountains and near the North and South Poles. In the past, during cold periods called Ice Ages, they were far more widespread, covering huge areas of North America and Europe. When the ice melted, dramatic marks were left on the landscape. In the mountains of northwest America and Scotland the ice gouged out giant, trough-like valleys. Over much of the American midwest and the plains of northern Europe lay vast deposits of till (rock debris).

Cracked ice

As a glacier moves downhill, it bends and stretches, opening deep cracks called crevasses. These may be covered by fresh falls of snow, making the glacier treacherous for climbers to cross. Crevasses may be a sign that the glacier is passing over a bar of rock on the valley floor. A bergschrund is a deep crevasse where the ice pulls away from the back wall of the cirque at the start of the glacier.

Old and new snow

The snow on these mountain peaks is likely to be quite different from that of the glacier below. Glaciers are made up of névé (new snow) and a compacted layer of old snow, or firn, beneath. All the air is squeezed out of the firn so that it looks like ice. The ice becomes more compacted over time, turning into thick white glacier ice, which begins to flow slowly downhill.

The course of a glacier

A glacier typically begins in a hollow high in the mountains called a cirque. It then spills out over the lip of the cirque and flows down the valley. If the underside of the glacier is "warm" (about 32°F), it glides in a big lump on a film of water melted by the pressure of ice. This is called basal slip. If the underside of the glacier is well below 32°F, it moves though there were layers within the ice slipping over each other like the shuffling of a deck of cards. This is called internal deformation. Glaciers usually move in this way high up where temperatures are lower, and by basal slip farther down.

Deep digging
The fjords of Norway were made by glaciers that carved out deep valleys well below the current sea level. When the ice retreated, sea flooded the valleys to form inlets that in places are over 3,200 ft deep. Glaciers may be slow, but their sheer weight and size gives them the immense power needed to mold the landscape. They carve out wide valleys, gouge great bowls out of mountains, and slice away entire hills and valleys as they move relentlessly on.

Moraine and drift
The gray bank across the picture above is a terminal (end) moraine. This is where debris has piled up in front of a glacier that has melted. The intense cold around a glacier causes rocks to shatter, and as the ice bulldozes through valleys, it shears huge amounts of rock from the valley walls. The glacier carries all this debris and drops it in piles called moraines. Melting glaciers deposit blankets of fine debris called glaciofluvial (ice river) drift.

Alaskan tundra
This landscape in Alaska is shaped by its periglacial climate (the climate near a glacial region). Winters are long and cold with temperatures always below freezing. In the short summers, ice melts only on the surface, and so the ground beneath is permafrost (permanently frozen). Water collects on the surface and makes the land boggy. As the ice melts, it stirs and buckles the ground beneath. As the ground thaws, then freezes again, cracks form, creating deep wedges of ice.

Glaciated valley
A wide, U-shaped valley in Scotland is left over from the Ice Ages, when glaciers were much more widespread. The last of the Ice Ages ended about 10,000 years ago. Valleys like this were carved by glaciers over tens of thousands of years. They are very different from the winding V-shape of a valley cut by a river.

DESERT LANDSCAPES

NOT ALL deserts are vast seas of sand. Some are rock-strewn plains. Others are huge blocks of mountains standing alone in wide basins, or just empty expanses of ice, as in Antarctica. All have the one common characteristic of being very dry. The lack of water makes desert landscapes very different from any others. Wind plays a much more important part in shaping them, because there is neither moisture to bind things together nor running water to mold them. In the desert, wind carves many unusual and unique landforms, from sculpted rocks to moving sand dunes. Very few places in the world are entirely without water, and intermittent (occasional) floods do have a dramatic effect on many desert landscapes. Instead of the rounded contours of wetter landscapes, steep cliffs, narrow gorges and pillar-like plateaus called mesas and buttes are formed.

Water in the desert
A valuable pocket of moisture has formed in the desert. There is often water beneath the surface of a desert, which may be left over from wetter days in the past. It may be water from wetter regions farther away, which has run down sloping rock layers beneath the desert. Occasionally, the fierce desert wind blows a hollow out of the sand so deep that it exposes this underground water.

Sand dune styles
In some deserts, such as the Sahara, there are vast seas of sand, called ergs, where the wind piles sand up into dunes. The type of dune depends on the amount of sand and the wind direction. Crescent-shaped dunes called parabolic dunes are common on coasts. Ones with tails facing away from the wind are called barchans. These dunes creep slowly forward. Transverse dunes form at right angles to the main wind direction where there is lots of sand. Seifs form when there is little sand and the wind comes from different directions.

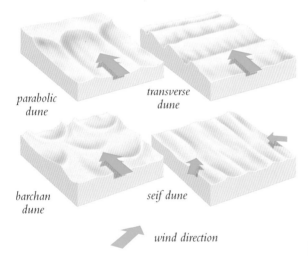

parabolic dune

transverse dune

barchan dune

seif dune

wind direction

Dune sea
In the western Sahara, giant sand dunes hundreds of yards high have formed as a result of two million years of dry conditions. In places, the wind has piled up the dunes into long ridges called draa, that stretch far across the desert like giant waves in the sea. Each year, these ridges are moved 80 ft farther by the wind, as sand is blown up one side and rolls down the far side. Space probes have seen dunes like these on the planet Mars.

Monuments to water

Monument Valley in Utah, is a spectacular example of what water erosion can do in dry places. The monuments are physical features called mesas. They are protected from water erosion by a cap of hard rock. The softer, unprotected rock between them has been washed away over millions of years. As water flows in channels rather than overland, there is nothing to round the contours, and cliffs remain sheer.

Occasional rivers

Rain is rare in the desert. What rain there is flows straight off the land and does not soak in, so streams rarely flow all the time. Instead, most streams flow only every now and then, and are said to be ephemeral or intermittent. In between wet periods they leave behind dry beds called arroyos. In the Sahara Desert and the Middle East, rare rain torrents wash out narrow gorges called wadis. These are normally dry, but after rain may fill rapidly with water in a flash flood.

The power of the wind

Strong winds blow unobstructed across the desert, picking up grains of sand and hurling them at rocks. The sandblasting can sculpt rocks into fantastic shapes such as this rock arch. Satellite pictures have revealed parallel rows of huge, wind-sculpted ridges in the Atacama Desert in Chile and in the Sahara Desert in Africa. These are called yardangs and are hundreds of yards high and dozens of miles long.

FACT BOX

- Summer temperatures in the Sudan Desert in Africa can soar to 133°F, hotter than anywhere else in the world.

- Because desert skies are clear, heat escapes at night and temperatures can be very low.

- Not all the world's deserts are hot. Among the world's biggest deserts are the Arctic and Antarctic, both of which have hardly more rain than the Sahara.

SEA BATTLES

COASTLINES ARE constantly changing shape. They change every second, as a new wave rolls in and drops back again, and every six hours, as the tide rises and falls. Over longer periods, too, coastlines are reshaped by the continuous assault of waves. They change more rapidly than any other type of landscape. Of all the agents that erode (wear away) the land, the sea is the most powerful of all.

Huge cliffs are carved out of mountains, broad platforms are sliced back through the toughest of rocks, and houses are left dangling over the edges of the land. Such examples are proof of the awesome force of waves. The sea is not always destructive, though. It also builds. Where headland cliffs are being eroded by waves, the bays between may fill with sand—often with the same material from the headlands. On low coasts where the sea is shallow, waves build beaches and banks of shingle, sand and mud. How a coastline shapes up depends on the sort of material it is made of and on the direction and power of the waves.

White cliffs
Waves can quickly wear the land into sheer cliffs like those at Beachy Head in England. When the last Ice Age ended about 8,000 years ago, the sea level rose as the ice melted. Sea flooded over the land to form what is now the English Channel. The waves quickly began to slice away the land. The valleys were cut off so that they became just dips in the cliffline, while the hills became crests.

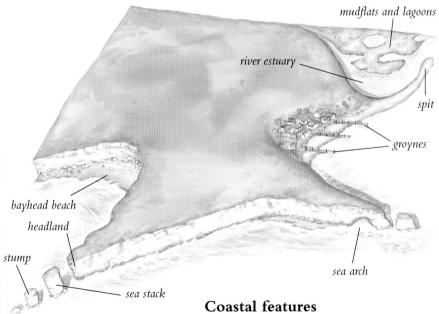

Coastal features
The sea's power to build and destroy a coastline can be seen in this picture. Exposed parts of the coast that face into the waves are eroded into steep cliffs. Headlands are worn back, leaving behind stacks, stumps and arches. In more sheltered places, the sand piles up to form beaches, or waves may carry material along the coast to build spits and mudflats.

Saving the beach
Low fences called groynes have been built along this beach to stop it from being carried away by the wave action. When waves hit a coast at an angle, they fall back down the beach at right angles. Any sand and shingle carried by a wave falls back slightly farther along the beach. In this way, sand and shingle is carried along the beach in a zigzag movement called longshore drift.

Storm force

During a storm, the waves crash onto the shore with tremendous force. The waves attack hard rock in two ways. They pound the rock with a huge weight of water filled with stones. They also split the rock apart as the waves force air into cracks. On high coasts, the constant attack of waves undercuts the foot of the slope, and unsupported upper parts topple down to create a cliff.

Wave-cut platforms

A pool has formed in a dip on a rocky platform on the seashore. The sea's erosive power is concentrated in a narrow band at the height of the waves. As the waves wear back sea cliffs, they leave the rock below wave height untouched. As the cliff retreats, the waves slice off a broad platform of rock. Geologists call this a wave-cut platform, and it lies between the low-tide and high-tide marks. As the tide goes out twice each day, the sea leaves water in dips and hollows to form pools.

Rock arch

The sea arch at Durdle Door in Dorset, England, is made by waves eating away at large blocks of well-jointed rock. The waves have worked their way into joints in the rock and slowly enlarged them. Eventually, the cracks are so big that they open up into sea caves, or cut right through the foot of a headland to create a sea arch like this. When further erosion makes the top of an arch collapse, pillars called stacks are left behind. Pillars may then be eroded into shorter stumps.

PROJECT

PULLING TIDES

Every 12 hours or so, the sea rises a little in some places, then falls back again. These rises and falls are called tides, and they are caused mostly by the Moon. The Moon is far away, but gravity pulls the Earth and Moon together quite strongly. The pull is enough to pull the water in the oceans into an egg shape around the Earth. This creates a bulge of water—a high tide—on each side of the world. As the Earth turns around, these bulges of water stay in the same place beneath the Moon. The effect is that they run around the world, making the tide rise and fall twice a day as each bulge passes. Actually, the continents get in the way of these tidal bulges, making the water move around in a complicated way.

The first experiment on these pages shows how the oceans can rise and fall a huge distance through tides without any change in the amount of water in the oceans at all. The second shows what the tidal bulge would look like if you could slice through the Earth, and how it moves around as the Earth turns beneath the Moon.

Low tide
The height of tides varies a great deal from place to place. In the open ocean, the water may not go up and down more than a yard or so. But in certain narrow inlets and enclosed seas, the water can bounce around until tides of 50 ft or more can build up.

HIGH AND LOW TIDE

You will need: round plastic bowl, water, big plastic ball to represent the world.

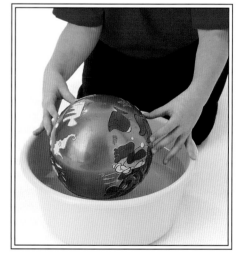

1. Place the bowl on a firm surface, then half fill it with water. Place the ball gently in the water so that it floats in the middle of the bowl.

2. Put both hands on the top of the ball, and push it down into the water gently but firmly. Look what happens to the level of water; it rises in a "high tide."

3. Let the ball gently rise again. Now you can see the water in the bowl dropping again. So the tide has risen and fallen, even though the amount of water is unchanged.

PROJECT

THE TIDAL BULGE

You will need: *strong glue, one 8-in. length and two 6-in. lengths of thin string, big plastic ball to represent the world, round plastic bowl, water, adult with a simple hand drill.*

1 Glue the 8-in. length of string very firmly to the ball and let dry. Ask an adult to drill two holes in the rim of the bowl on opposite sides.

2 Thread a 16-in. length of string through each hole and knot around the rim. Half fill the plastic bowl with water and float the ball in the water.

3 Ask a friend to pull the string on the ball toward him or her. There is now more water on one side of the ball. This is a tidal bulge.

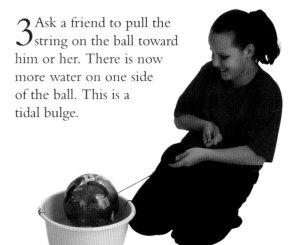

Pulling power

When the Moon and Sun line up at a Full Moon and a New Moon, their pulling power combines to create very high spring tides. A Half Moon means that the high tide will fall well below the highest tide mark. This is because the Sun and the Moon are at right angles to each other. Even though the Sun is farther away from Earth than the Moon, it is so big that its gravity still has a tidal effect. But at a Half Moon, they work against each other and create the very shallow neap tides.

4 The Moon pulls on the water as well as the Earth. So now ask the friend to hold the ball in place while both of you pull out the strings attached to the bowl until it distorts.

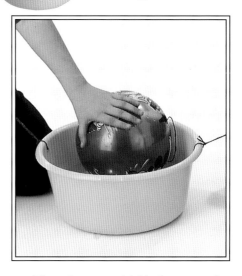

5 There is now a tidal bulge on each side of the world. One of you slowly turn the ball. Now you can see how, in effect, the tidal bulges move around the world as the world turns.

101

THE OCEANS' HIDDEN DEPTHS

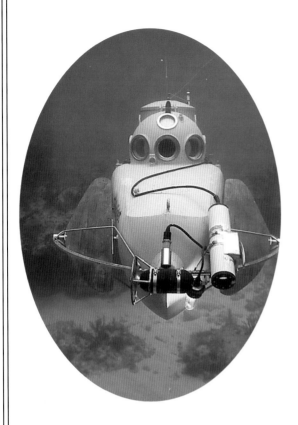

NEARLY THREE-QUARTERS of the world is under the oceans, lying an average of 12,238 feet underwater. In places, the oceans plunge down to a depth of 36,000 feet, which is enough to drown the world's highest mountain, Mount Everest, and leave more than a mile to spare. Seas cover four-fifths of the Southern Hemisphere, and three-fifths of the Northern Hemisphere. The five great oceans around the world are the Pacific, the Atlantic, the Indian, the Southern (around Antarctica) and the Arctic. The biggest of them by far (although they are all actually linked together) is the Pacific, which covers almost a third of the Earth. Until quite recently we knew little more about the ocean depths than about the surface of Mars. However, in the last 40 years, there have been remarkable voyages in submersible craft capable of plunging to ever greater depths, and extensive oceanographic (ocean mapping) surveys. These have revealed an undersea landscape as varied as the continents, with mountains, plains and valleys.

Probing the depths

Knowledge of the ocean depths has increased dramatically, thanks to small, titanium-skinned submersibles and robot ROVs (remote-operated vehicles). These can withstand the enormous pressure of 1¾ mi. of water above them. Satellites orbiting high above the Earth can make instant maps of the sea floor too. They pick up faint variations in the sea surface. These are created by changes in gravity, which in turn are caused by ups and downs in the ocean floor.

FACT BOX

• Seawater is 96.5% water and 3.5% salt. Most of the salt is sodium chloride (table salt).

• There are canyons under the sea as big as the Grand Canyon.

• The Mid-ocean Ridge is the world's longest mountain chain, winding 23,000 mi. under three oceans, including the Atlantic.

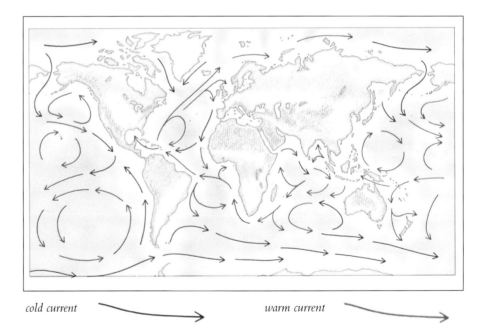

cold current ⟶ *warm current* ⟶

Currents on the ocean surface

The ocean waters are constantly circulating in currents. Those near the surface are driven along by the combined effect of winds and the Earth's rotation. They circulate in giant rings called gyres. In the Northern Hemisphere gyres flow clockwise, while in the Southern Hemisphere they flow counterclockwise. Deeper down, currents flow between the poles and the equator. These are driven by differences in the water density, which varies according to the temperature and how salty the water is.

Island rings
The Maldives are a series of atolls—ring-shaped islands of coral—in the Indian Ocean. The coral ring first began to form around the peak of a seabed volcano that poked up above the sea's surface. At some time, the seabed moved. The volcano moved with it and slowly sank beneath the waves. The coral, however, kept on growing upward, without its volcano center. The reef is sometimes hundreds of yards deep.

Coral reefs
Over millions of years huge colonies of tiny sea animals called coral polyps build reefs (ridges) just below the surface of the sea. As each polyp dies its skeleton becomes hard. Colorful living polyps live on the skeletons of dead ones, so gradually layers of polyps build up and the coral reef grows bigger. Coral reefs support an extraordinary variety of marine life.

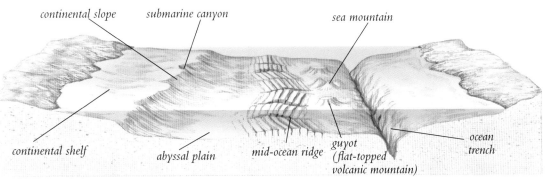

The ocean floor
Running along the edges of each continent is a narrow shelf of shallow ocean barely 300 ft deep, called the continental shelf. Beyond, the ocean bed plunges down the continental slope into the depths of the ocean basin 6,500 ft below. The floor of the ocean basin is an almost flat plain, called the abyssal plain, which is covered with a thick slime called ooze.

Giant waves
The world's biggest waves occur in the biggest oceans, the Pacific and the Atlantic—and in the Southern Ocean around Antarctica. The more winds blow over the oceans, the bigger waves are likely to be. In the Southern Ocean, the winds roar around the world unhindered by land. Monster waves estimated to be 130 to 160 ft high—as high as an apartment building—have been spotted on occasion.

PROJECT

OCEANS ON THE MOVE

You will need: *rectangular tub, pitcher, water, bathtub or inflatable wading pool.*

THE OCEANS are very rarely completely still. Even on the calmest day, little ripples play across the surface, or the water gently undulates. When the weather is stormy, giant waves higher than a house can rear up and crash down, turning the sea into a raging turmoil.

Waves begin as the surface of the water is whipped up into little ripples by wind blowing across the surface. If the wind is strong enough and blows far enough, the ripples build up into waves. The stronger the wind and the longer the fetch (the farther they blow across the water), the bigger the waves become. In big oceans, the fetch is so huge that smooth, giant, regular waves called swells sweep across the surface, and waves may travel thousands of miles before they meet land.

Waves usually only affect the surface of the water. The water does move at a deeper level, in giant streams called ocean currents, if the wind blows again and again from the same direction. Some deep ocean currents, moved by differences in the water's saltiness or temperature, can stir up the water right down to the ocean bed. The first project shows how waves are made. Currents such as the ones in the second project happen in the oceans on a much larger scale, and circulations or gyres such as this swirl around all the world's major oceans.

MAKING WAVES

1 Place the tub on the floor or on a table. Choose a place where it does not matter if a little water spills. Fill the tub with water almost to the top.

2 Blow very gently on the surface of the water. You will see that the water begins to ripple where you blow on it. This is how waves are formed by air movement.

3 Fill the bathtub or pool with water. Blow gently along the length of the bathtub or pool. Blow at the same strength as in step 2, and from the same height above the water.

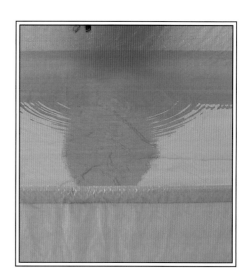

4 Keep blowing for a minute or so. Notice that the waves are bigger in the bath or pool, even though you are not blowing harder. This is because the fetch is bigger.

PROJECT

OCEAN CURRENTS

You will need: *rectangular tub, pitcher, water, talcum powder.*

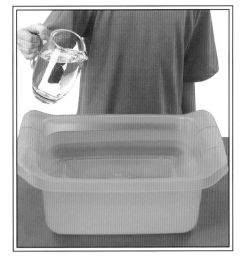

1. Place the tub on the floor or on a table. Choose a place where it does not matter if a little water spills. Fill the tub with water almost to the top.

2. Sprinkle a small amount of talcum powder on the water. Use just enough powder to make a very fine film on the surface. The less you use, the better.

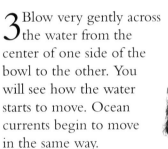

Wind and spin

Waves in the Antarctic oceans are driven by the wind, ocean surface and the effect of the Earth's rotation. The main winds in the tropics, called trade winds, blow the waters westward along the equator, creating equatorial currents. When these currents run into continents, they are deflected, still warm, toward the poles by the Earth's rotation. Eventually, they run into westerly winds that blow the water back eastward again.

3. Blow very gently across the water from the center of one side of the bowl to the other. You will see how the water starts to move. Ocean currents begin to move in the same way.

4. Keep blowing, and the powder will swirl in two circles as it hits the far side. This is what happens when currents hit continents. One current turns clockwise, the other turns counterclockwise.

THE WORLD'S WEATHER

Storms, winds, snow, rain, sunshine and all the other things we call weather are simply changes in the air. Sometimes these changes can happen very suddenly. A warm sunny day can turn into a stormy one, bringing high winds and torrents of rain that then end just as abruptly as they began. In some places, such as in the tropics, on either side of the equator, there is very little difference in the weather from one day to another.

The planet's weather and every change in it is governed by the heat of the Sun. Winds stir up, for example, when the Sun heats some places more than others. This sets the air moving. Rain falls when air warmed by the Sun lifts moisture high enough for it to condense into big drops of water. On satellite photographs, swirls of cloud indicate how the air is moving. From them, meteorologists can identify distinct circulation patterns and weather systems, such as depressions and fronts, each of which brings a particular kind of weather.

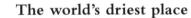

Antarctic cold
The coldest place in the world is Antarctica. In Vostok, the Russian research station in Antarctica, the temperature averages −72°F, and once dropped to −126.4°F. The Sun strikes polar regions at a low angle, not from directly overhead as it does at the equator, so its power is severely reduced. In winter, the Sun is below the horizon for most of the time, and it is night in the polar region for three icy months.

The world's driest place
The Atacama Desert in Chile is the world's driest place, receiving little more than ½ in. of rain in a year. It is dry because winds blow in from the Pacific Ocean over cold coastal currents. The cold water cools the air so much that all the moisture in it condenses before it reaches the land. So, as it blows over the Atacama, it is very dry.

Winds of the world
Some winds are local, others blow only for a short while. Prevailing winds are those that blow for much of the year. The map shows the world's three major belts of prevailing winds. Trade winds blow between the Tropics of Cancer and Capricorn on either side of the equator. They are dry winds from the east. Moist westerlies from the west blow between the tropics and the polar regions. Icy polar easterlies blow around the North and South Poles.

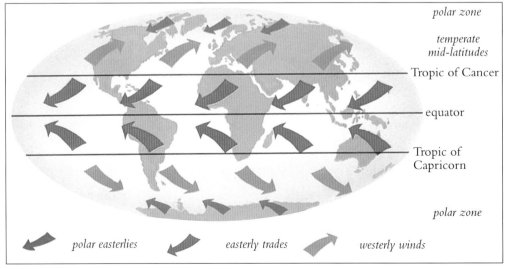

Green and moist

The pasturelands of England are lush and green because they are well-watered by rain. England is in the westerly wind belt. Most of the winds blow in from the west over the Atlantic Ocean, where they pick up plenty of moisture that later falls as rain. Westerly winds also bring storms called depressions. These are places where winds spiral in toward a core of low pressure air. Depressions bring rain storms to the west coasts of Europe and North America as they move slowly east.

Perfect climate

Few places have such perfect weather as Quito in South America. It is close to the equator, so the air is warm, but because it is high up, it never becomes too warm. The temperature in Quito never drops below 46°F at night or rises above 72°F during the day. The weather is made even more perfect by the fact that just 100 mm of rain falls each month. It is no wonder, then, that the city of Quito is called the "Land of Eternal Spring."

FACT BOX

- The wettest place in the world is Tutunendo in Colombia, where the rainfall averages 460 in. in a year.

- The hottest place is Dallol in Ethiopia, where it averages 94°F in the shade.

- The place with the most extreme weather is Yakutsk in Siberia. Here winters can be –83.2°F and summers 102°F.

Monsoon rains

For six months of the year, some parts of the tropics, such as India, are parched dry. Then, suddenly, torrential monsoon rains arrive, as the summer Sun heats the land. Air warmed by the land rises, and cool, moist air from the sea is drawn in underneath. This rain-bearing air pushes inland. Showers of heavy rain pour down during the wet season. Then, after about six months, the land cools and the winds reverse and blow out to sea. Immediately, the rain eases and the dry season is back.

CLIMATE CHANGE

Storms warn of global warming
Extra warmth from global warming puts extra energy into the air, bringing storms as well as warmer weather. The industrial world pumps huge amounts of gases into the air, including carbon dioxide from burning oil in cars and power stations. These greenhouse gases are so called because they trap the Sun's heat in the atmosphere like the glass in a greenhouse.

EIGHTEEN THOUSAND years ago, the world was bitterly cold. A third of the planet was covered by thick sheets of ice. Vast glaciers spread over much of Europe and pushed far south into North America. This was just the most recent Ice Age. In the future, there will be another. The world's climate changes constantly, becoming warmer or colder from one year to the next, by the century or over thousands of years. These changes may be caused by a shift in the Earth's position relative to the Sun, or by bursts of sunspot activity in the Sun. Natural events such as volcanic eruptions, the impact of meteorites, and the movement of continents, also affect the weather. Recently, scientists have been concerned by the sudden warming of the world, triggered by air pollution.

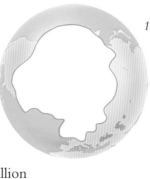

Sunspot storms
Sunspots are dark spots on the Sun where the surface is less hot. They seem to change constantly and reach a peak every 11 years or so. Measurements from the Nimbus-9 satellite show that the Sun gives the Earth less heat when there are fewer sunspots. Weather records show that when they reach their maximum level, the weather on Earth is warmer and stormier.

Ice core
A scientist investigates a core of solid ice that provides a remarkable record of climate change. The polar ice caps were built up over hundreds of thousands of years. Scientists drill into the ice of Greenland and Antarctica, and extract ice cores that are made up of layers of snow that have fallen over the years. They can detect changes in the atmosphere from microscopic bubbles of ancient air trapped within the ice and see how greenhouse gases have increased.

Polar ice caps
During Ice Ages, Earth becomes so cold that the polar ice caps grow to cover nearby continents with vast sheets of ice. Ice Ages are periods of time lasting for millions of years. There have been four in the last billion years. During an Ice Age, weather varies from cold to warm over thousands of years, and ice comes and goes. There have been 17 glacials (cold periods) and interglacials (warmer periods) in the last 1.6 million years.

polar ice cap 18,000 years ago

polar ice cap today

Moving land

Everywhere on Earth has had a very different climate at some point in history. Fossils show that New York was once a desert, and icy Antarctica once enjoyed a tropical climate. This fossil is of a tropical fern, but it was found in Spitsbergen, which is well inside the Arctic Circle. Corals only survive in warm seas, but strangely have been found in cold, northern seas. Such dramatic differences are not due to changes in the global climate but because the continents have drifted around the globe.

Antarctic ice

The amount of ice in the world is always fluctuating. Antarctica contains 95 percent of the world's ice and snow. But even Antarctica has not always been covered in ice. In fact, most Antarctic ice is less than ten million years old. Icicles form as the warmer weather comes and the ice begins to melt.

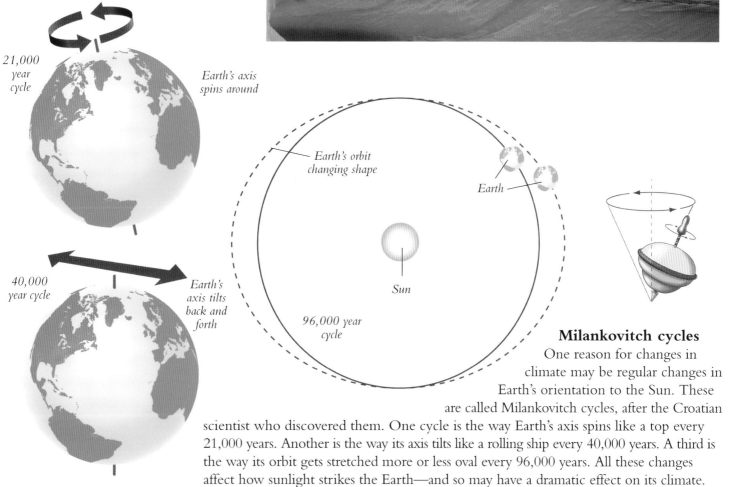

Milankovitch cycles

One reason for changes in climate may be regular changes in Earth's orientation to the Sun. These are called Milankovitch cycles, after the Croatian scientist who discovered them. One cycle is the way Earth's axis spins like a top every 21,000 years. Another is the way its axis tilts like a rolling ship every 40,000 years. A third is the way its orbit gets stretched more or less oval every 96,000 years. All these changes affect how sunlight strikes the Earth—and so may have a dramatic effect on its climate.

PROJECT

WEATHER RECORD

Ancient wood
You may find a fallen or freshly sawn tree in a forest to study. Scientists can take a record from a living tree by drilling out a small rod through the tree with a device called an increment borer. This gives a detailed record of climate changes Studying the past through tree rings is called dendrochronology, from the Greek words *dendron* (tree) and *chronos* (time).

CLIMATE IS the word used to describe the typical weather of a place at a particular time. The world's climates have changed throughout history. There have been times in the periods between Ice Ages, for example, when some plants and animals that are found in hot lands today lived in more northerly regions than they do now.

We can see how the climate has changed by studying weather records that have been made by people in the past, but these rarely go back more than 200 years. Scientists can also find many clues to climate changes in nature. In the sediments of sand and mud on seabeds, for example, they found fossils of tiny shellfish called *Globorotalia*, which coils to the left in cold water and to the right in warm water. By figuring out when the sediments were laid down, the scientists could discover whether the water was warm or cold by the way the shellfish were lying.

This project shows how to make your own discoveries about recent climate change by looking at the year-by-year record of tree growth that is preserved in wood.

THE WOODEN WEATHER RECORD

You will need: *newly cut log, decorator's paintbrush, ruler with millimeter measurements, metric graph paper, pencil, calculator.*

1 Ask a tree surgeon or a sawmill for a newly cut slice of log. Use the paintbrush to brush off the dust and dirt from the slice of wood.

2 When the log slice is clean, examine it closely. Look at the pattern of rings. They are small in the center and get bigger and bigger toward the outer edge of the log.

PROJECT

FACT BOX

• Each ring in the tree's cross-section represents a year's growth.

• The strong line at the edge of each ring marks the time in winter when growth stops.

• A wide ring indicates a warm summer with good growth.

• A narrow ring indicates a cool summer with poor growth.

• See if you can spot good summers and bad.

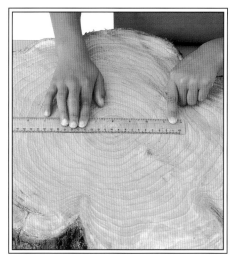

3 Each ring is a year's growth. So count the rings out from the center carefully. This tells you how old the tree is. If there are 105 rings, for example, the tree is 105 years old.

4 Using a ruler, measure the width of each ring. Start from the center and work outward. Ask a friend to write down the widths as you call them out.

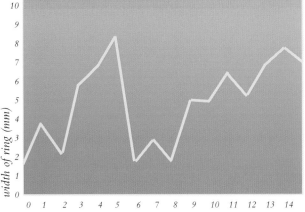

Long-term calendar

All kinds of trees are useful for tree ring analysis, though the sequoias (redwoods) and pines of California are especially valued because some trees are over 4,000 years old. By comparing rings from different trees, scientists can build up a record of climate change.

5 On graph paper, mark the years along the bottom line, using five small squares for each year. Mark widths for the rings up the side of the paper, using five small squares for each millimeter. Now, starting with year one on the left, plot your measurements as dots for each year across the graph.

6 Connect the dots with a pencil line. This line shows how the weather has changed with each year. If the line is going up, the weather was warmer. If the line falls, the weather was colder. See if you can spot if it is getting warmer or colder over time.

VEGETATION ZONES

SOME PLANTS, such as alpine grasses, can survive in very cold conditions, even if they are covered with thick snow for several months. Others, such as cacti, can cope with extreme heat. Each kind of plant thrives under particular conditions of soil and climate. Some groups of plants are so well adapted to conditions that exist in a particular region of the world that they are identified with those regions. The world can therefore be split into plant or vegetation regions according to the kind of plants that thrive there. Climate is the biggest influence on the kind of plants that grow, so vegetation regions tend to coincide with regions that have particular climates, such as tropical (near the equator), or polar (near the poles). Many different plants live in each place, and within these broad regions, conditions can vary enormously.

Barren tundra
Only lichens, mosses, hardy grasses and tiny shrubs such as dwarf willows and birches grow in the tundra wastelands. In these polar regions, the temperature rarely rises above freezing, and then only in a few months of the year. Plants must survive on little or no water in winter because it is frozen. Then, when the ice melts in spring, they have to cope with ground that is completely waterlogged.

Northern forests
Across the north of Russia and Canada are vast coniferous forests. These vegetation zones are called boreal forests or taigas. Winters are dark and cold, with thick snow. Conifers have thin, needle-like leaves that resist the cold, and snow falls easily off the cone-shaped trees.

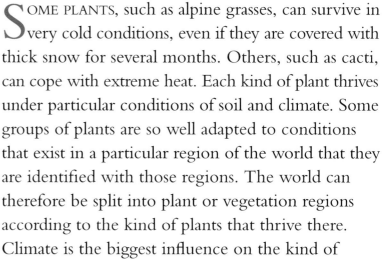

Temperate grassland
Vast areas of the temperate zone, between the tropics and the poles, are covered with grass. Some of this has been created by farmers, as they have cleared woods for pasture, but much is natural. Temperate grassland is called by different names in different parts of the world, such as steppe in Asia and prairie in North America. Steppes are dry, so the grass is very short and coarse. Prairies are damper and the grass is lusher and longer.

Mixed woodland
In temperate regions, where summers are warm and quite moist, but winters are cool, the native vegetation is deciduous woodland. Deciduous trees lose their leaves in autumn. This reduces the need for water in winter, when frozen ground limits the water supply. Much of Europe and North America was once covered by vast deciduous woods, but over the centuries trees have been cut down or burned to make way for farmland.

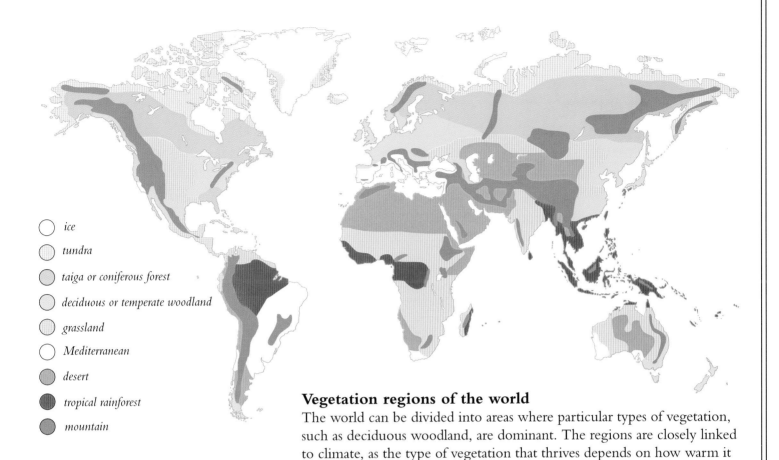

- ○ ice
- ○ tundra
- ○ taiga or coniferous forest
- ○ deciduous or temperate woodland
- ○ grassland
- ○ Mediterranean
- ○ desert
- ○ tropical rainforest
- ○ mountain

Vegetation regions of the world

The world can be divided into areas where particular types of vegetation, such as deciduous woodland, are dominant. The regions are closely linked to climate, as the type of vegetation that thrives depends on how warm it is or whether there is rain throughout the year or only in particular seasons. As the climate becomes colder away from the equator toward the poles (and higher up mountains), there are fewer different types of plants.

Tropical rainforest

Warmth and plentiful rain throughout the year make tropical rainforests the richest plant habitats on Earth. Deciduous woods rarely have more than a dozen tree species, but tropical rainforests may have 100 or more in a single hectare. The forests are surprisingly fragile because trees and soil are dependent on each other for survival. If trees are cut down, soil and the nourishing things it contains are quickly washed away.

Tropical grassland

Where rain in the tropics is seasonal, trees are rare, as they cannot cope with the long dry season. The typical vegetation is grassland, which in Africa is called savanna. Grasses in savanna lands grow tall and stiff. Dark evergreen trees such as acacias survive because their waxy leaves retain water, and their thorns protect them from animals in search of moisture.

THE BALANCE OF LIFE

LIFE ON Earth may be classified into thousands of ecosystems. These are communities of living things that interact with each other and with their surroundings. An ecosystem can be anything from a piece of rotting wood to a huge swamp, but all the living things within it depend on each other.

Each living thing also has its own favorite place where factors such as temperature and moisture are just right. Some species can survive in a variety of habitats, but many can cope only with one. In an ecosystem, organisms depend on each other, and taking away just one species can threaten the existence of the others. If the plants on which a certain caterpillar feeds are destroyed, for example, the caterpillar dies, the birds feeding on the caterpillar starve and the foxes that feed on the birds go hungry, too.

Underwater richness
Coral reefs are the rainforests of the oceans. They provide shelter and food for an enormous range of marine plants and animals, from tiny coral polyps to giant clams and vicious predators. It is a fierce battle for life, food and space, however, and each species must develop its own program for survival. Even the starfish and seasquirts in this picture are predators.

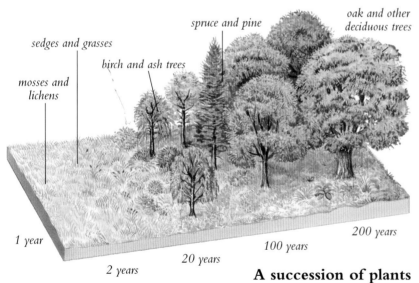

A succession of plants
When there is enough warmth and moisture on a piece of bare, rocky land, the first plants will grow. These will be the smallest, simplest plants, such as mosses and lichens, then tough grasses, that do not need very much to live on. The plants begin to hold the soil together. As they die and rot, they add nutrients to the soil, preparing it for bigger plants to grow. Soon there is enough to support small shrubs and tough trees such as pines, and eventually deciduous trees such as the oaks. This process is called vegetation succession, and it takes about 200 years for deciduous woodland to develop from the moss and lichen stage.

Harvest time
Farmland such as this has destroyed the natural vegetation and ecosystems. The number of plant species has dramatically reduced. A forest with hundreds of different plants may have been cleared to make way for a single crop. Because farming interrupts the flow of nutrients between soil and plants, the soils quickly become depleted, and farmers add artificial fertilizers.

Feeding habits

All animals depend on other living things for food and form part of an endless chain. This picture shows how the food chain, or web, works. A grasshopper eats a leaf, a thrush may eat the grasshopper and a kestrel may eat the thrush. When the kestrel dies and falls to the ground, bacteria break its body down and add nutrients to the soil so that new plants can grow. Herbivores eat only plants. Carnivores are meat eaters, and omnivores eat both vegetable and plant matter. Plants and algae make their own food from sunlight, and so are called autotrophs (self-feeders).

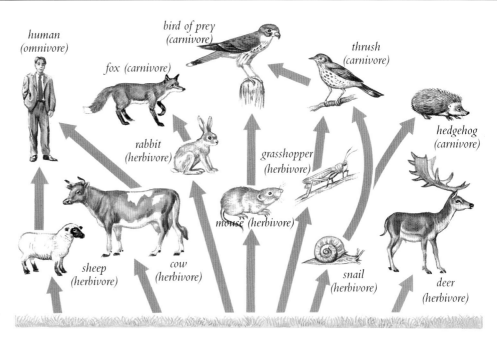

A wealth of natural life

Swamps, ponds and all the other places known together as wetlands, were once seen as useless land that could not be farmed or built on. More than half the wetlands in the United States have been drained in the last 100 years or so. However, wetlands are remarkably rich environments, producing up to eight times as much plant matter as the average wheat field. They can also play an important role in controlling floods and are a valuable source of water in times of drought.

Layered rainforest life

Different types of plant and animals inhabit different levels of the tropical forest. Isolated trees shoot up to emerge above the dense, leafy roof of the main forest. Some are 200 ft tall. Below is a dense canopy of leaves and branches on top of tall, straight trees, 100–160 ft tall. In the gloomy understory beneath, young trees and shrubs grow and clinging lianas (climbing plants) wind their way up the trees.

FACT BOX

- Tropical rainforests cover less than 8 percent of the Earth's land surface.

- They make up half of the world's growing wood and provide a home for 40 percent of plant and animal species.

PROJECT
LIVING TOGETHER

THE WHOLE living world is a vast and everchanging puzzle of plant and animal life. Each organism that is part of this living puzzle links or interacts with other living things, either directly or indirectly. The whole picture is so huge and complicated that even for scientists, it is difficult to understand how it all works together all at once.

To make sense of it all, ecologists often break the living world down into lots of smaller units, such as tropical rainforests or freshwater lakes. Then they might break it down further into smaller regions, such as a mountain slope. They might go further still to identify individual trees or a pool on a rocky seashore. Each of these units, where the things living there interact with each other, is called an ecosystem. One way in which plant and animal life interacts is through their food chain, which show what eats what in an ecosystem. Warmth and shelter and protection from predators are other ways in which plants and animals can benefit from each other by co-existing in an ecosystem.

Arrested life
The axolotl from the lakes near Mexico City never grows up into a tiger salamander. It stays a tadpole all its life. This is because the water it lives in lacks iodine, the vital ingredient to make it grow. If the axolotl is given iodine injections, it turns into a tiger salamander. But that would alter the whole balance of life in the lakes. This shows how delicate the relationship is between each living thing and its environment.

MAKING YOUR OWN AQUARIUM ECOSYSTEM

You will need: gravel, net, plastic bowl, water, pitcher, glass aquarium, rocks and lumps of old wood, water plants, pitcher full of pondwater, water animals.

1 Put the gravel in a net and rinse in a plastic bowl of water or run it under the coldwater in the sink. This will discourage the formation of green algae.

2 Spread the gravel unevenly on the base of the tank to a depth of about 1¼ in. Add rocks and pieces of wood. These provide surfaces for the snails to feed on.

PROJECT

3 Fill the tank to about the halfway mark with tap water. Pour the water gently from a pitcher to avoid disturbing the landscape and churning up the gravel.

4 Add some water plants from an pet store. Keep some of them in their pots, but take the others out gently. Then root them in the gravel.

5 Now add a pitcher full of pondwater. This will contain organisms such as *daphnia* (water fleas), which add to the life of your aquarium.

6 Now add a few water animals you have collected from local ponds, such as tadpoles in frog spawn or water snails. Take care not to overcrowd the aquarium.

Specialized living

This giant turtle lives on the Galapagos Islands. It was here that the ecologist Charles Darwin noticed how island plants and animals adapt to their local environment in quite distinctive ways because they are isolated. Although the turtles originally had no natural enemies, they are now threatened with extinction.

7 Place the tank in a reasonably bright light, but not in direct sunlight. You can watch the plants in the tank grow. Keep the water clean by removing dead matter from the gravel every 6 weeks.

HUMAN IMPACT

HUMANS NOW dominate the Earth to a greater extent than any other species of animal has ever done before. The Earth seems to be in grave danger of suffering irreparable harm from our activities. The demands that humans make on the planet so that they can feed themselves and live in comfort damage the atmosphere, the Earth itself and plant and animal life. Car exhaust and factory chimneys choke the air with pollution. Gases from supersonic jets and refrigerator factories make holes in the atmosphere's protective ozone layer. Rivers are poisoned by agricultural and industrial chemicals. Unique species of plants and animals vanish forever as their habitats are destroyed. Forests are felled, vast areas of countryside are buried under concrete and beautiful marine environments are destroyed by tourism and sea traffic. The problem is not new, but as the pace of economic development increases, it is becoming more and more urgent to halt the destruction.

From forest to wasteland
A hillside that was once rich tropical forest has been slashed, burned and bulldozed. Vast areas of rainforest are being destroyed, in Brazil and Indonesia especially, to provide wood and to clear the land for rearing cattle. Unprotected soil soon turns to dust in the tropical Sun, and farmers move on to wreak destruction on fresh forest.

Poisoned air
Cars, factories and homes pour fumes into the atmosphere and are making the air increasingly poisonous to humans and plants. Lead has been eliminated in car fuels because of the damage it was causing to children's brains, and other substances may be responsible for a rise in lung diseases. Burning fossil fuels add sulfur dioxide to the water vapor in the air and cause acid rain, which pollutes lakes and kills trees, like these on Smokey Mountain.

Deadly algae bloom
A choked and lifeless river such as this is common. Few rivers in the world are entirely free from pollution. Of 78 rivers tested in China, 54 were badly polluted with sewage and factory waste. In Europe, most rivers have high levels of nitrates and phosphates from chemical fertilizers washed off farmland. Heavily manured land can make nearby streams so rich in organic matter that algae multiplies and chokes all other life.

Lifesaver
The rosy periwinkle is a tiny plant native to Madagascar. It was found to contain a chemical that has raised the chances of children surviving leukemia from 10 to 95 percent, by preventing cell division. Each hour, about 2,400 hectares of the world's rainforests are destroyed. Much will include precious plants like this whose value we will never know.

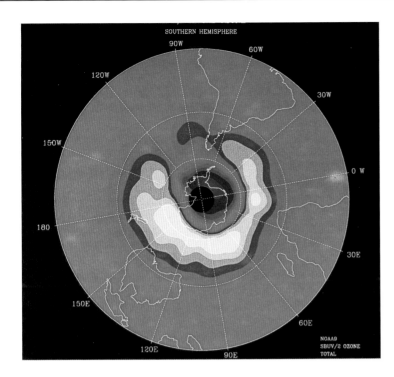

Destroying natural sunblock
The yellow and green blob on this satellite image of Antarctica is a huge hole in the ozone layer. Life on Earth depends on tiny amounts of the gas ozone in the stratosphere, 7½ to 31 mi. above the Earth. The ozone layer is our natural sunblock, and without it we have nothing to shield us from deadly ultraviolet (UV) rays from the Sun. This ozone hole reappears every spring, at both poles, each time getting bigger and staying for longer.

Exhaust fumes
Motor-vehicle exhaust discharges a wide range of unpleasant chemicals. Unburned hydrocarbons (better known as soot) make everything dirty and can cause breathing problems, while carbon dioxide adds to the greenhouse effect. Secondary pollutants are formed when exhaust mixes in the air. The worst of these may be ozone. This is a good sunblock in the atmosphere, but dangerous when inhaled. Soot reacts with sunlight and causes ozone-thick smog.

Endangered
The snow leopard is just one of millions of animal and plant species threatened with extinction by hunting or loss of natural habitat. Many species became extinct naturally, as the climate changed or a food source ran out. Today the rate of extinction is 400 times faster than the all-time average, all due to human interference.

The greenhouse effect
Carbon dioxide in the air is important because it helps to trap warmth from the Sun, like the panes of glass in a greenhouse. In the past, this greenhouse effect has kept the Earth nicely warm. However, burning fossil fuels such as coal and oil have increased levels of carbon dioxide dramatically, and this is collecting around the Earth. Vital waves of heat radiated from the Sun can filter through this layer to warm the Earth. But heat waves generated on Earth are becoming trapped. They hit the carbon dioxide barrier and bounce back again. This is making the Earth warmer. Experts think temperatures will go up 39°F in the next 100 years, bringing extremes of weather, rising sea levels and flooding.

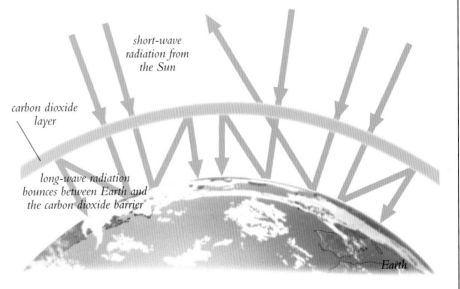

DANCE OF THE CONTINENTS

ONE OF the most amazing scientific discoveries of the 1900s was the idea of continental drift. Scientists discovered that the world's continents are not set in one place but are drifting slowly around the world—sometimes meeting, sometimes breaking apart. Recent high-precision measurements by satellite show that the continents are moving even now at between ¾ and 8 inches a year—about the pace of a fingernail growing. This may seem slow, but over the hundreds of millions of years of Earth's history, the continents have moved huge distances. There are ancient magnetic rocks within them that are like frozen compasses. Scientists can use them to plot how the alignment of the rocks has changed, and how the continents have twisted and turned. Piecing together these and other clues has gradually revealed just how the continents have moved over the last 750 million years.

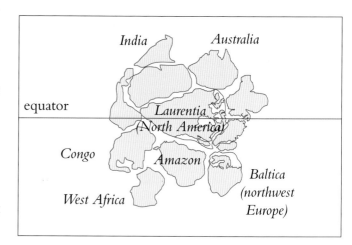

700 million years ago (mya)
All the continents are welded together in one giant continent that today's geologists call Rodinia. There are none of the recognizable shapes of today's continents. Magnetic clues in the rocks confirm that North America lay at the continent's heart along the equator and northwest Europe to the south.

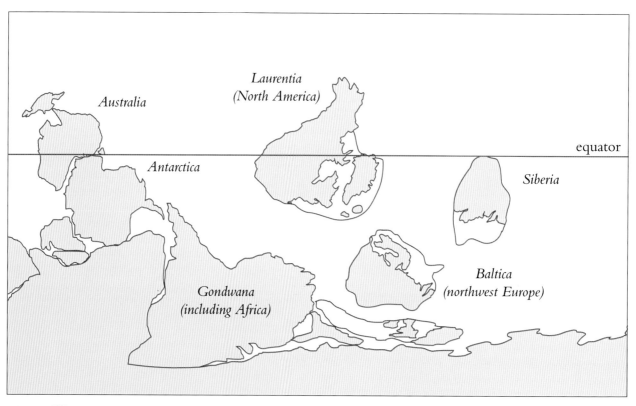

500 million years ago
The simple continent of Rodinia has broken up, but some of the fragments have gathered again around the South Pole. The map projection exaggerates the size of this South Pole continent, called Gondwanaland, but it was still massive, including all of today's Antarctica, Australia, South America, Africa and India.

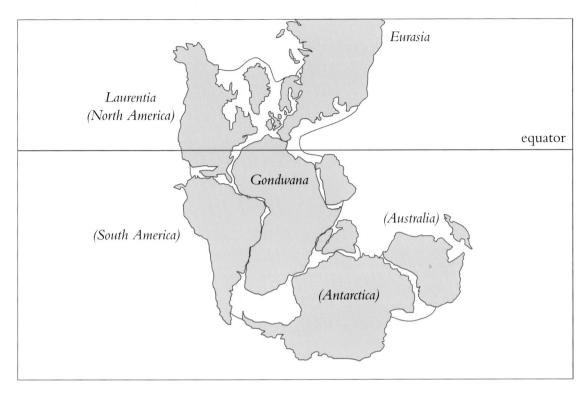

250 million years ago
Gondwanaland has merged with Laurentia and all the other continental fragments. A single mega-continent has formed and sits astride the equator. Geologists call this super-continent, Pangea (all earth). Pangea was surrounded by a single ocean, which geologists call Panthalassa (all sea). 200 mya, soon after the dinosaurs first appeared on Earth, Pangea began to break up.

50 million years ago
Between 200 and 50 mya, Pangea slowly broke up. First the Tethys Sea between Eurasia and Africa was opened up. Then the land split apart between Africa and South America to open into the South Atlantic ocean. By 50 mya, North America had drifted away from Europe to open up the North Atlantic. India was powering north into southern Asia. Australia was out on its own.

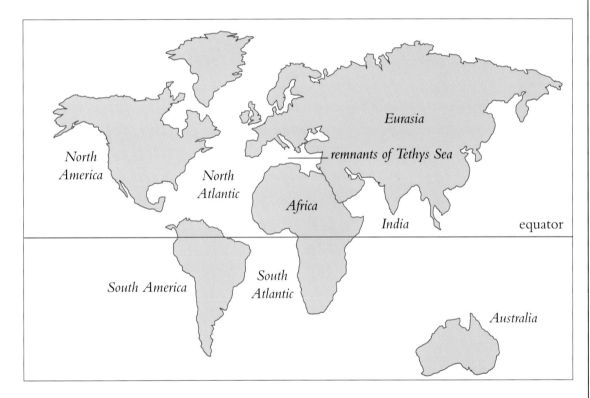

The world today
The continents today look familiar and set. But they are moving even now. In another 100 million years time the map of the world will look very, very different. The Americas are moving so far west that they will probably bump into eastern Asia in time, obliterating the Pacific. Africa will split into two parts and its eastern side will drift into southern Asia. As for the rest of Earth, only time will tell.

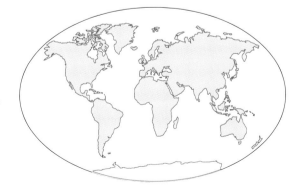

GLOSSARY

A

abyssal plain
The lowest level of the seabed, at a depth of up to 16,400ft down.

acid
A class of chemical compounds that contain the element hydrogen combined with other elements. Hydrogen splits from the other elements when acid is dissolved in water.

acid rain
Type of rain created when pollution by sulphur dioxide and nitrogen dioxide reacts in sunlight with oxygen and moisture in the air.

alluvial
Anything to do with sand and silt deposited by rivers. An alluvial plain is built up from silt deposited by a river.

amber
A pale brown or yellow transparent semiprecious stone formed by fossilized tree resin. It sometimes contains beautifully preserved prehistoric insects that became trapped in the sticky resin.

ammonite
A sea creature, now extinct, related to modern squids. It lived in a coiled shell, now found fossilized in many sedimentary rocks.

anticline
An arched upfold in layers of rock.

anthracite
A variety of hard, shiny coal that burns with hardly any smoke and gives more heat than any other kind of coal.

archebacteria
A very ancient form of bacteria able to survive in extreme conditions, for example in scalding volcanic waters in black smokers on the sea floor. They may have been the first life-forms on Earth.

arete
A sharp, knife-edge ridge between two mountain glaciers.

atmosphere
The blanket of gases surrounding the Earth.

atoll
A ring-shaped coral island created as the reef around a volcanic peak goes on growing at sea level while the volcano sinks beneath the waves.

atom
The smallest part of a chemical element. It is made up of many small particles, including electrons, neutrons and protons.

auroras
Spectacular displays of colored lights in the night sky above the North and South poles.

B

barchan
Crescent-shaped dune in a sandy desert.

base level
The lowest level a river can wear down to – usually sea level.

berghschrund
A deep crevasse at the head of a glacier, made as the ice pulls away from the valley side.

biome
A community of plants and animals adapted to similar conditions over large regions of the world. Sometimes called major life zones.

black smoker
Chimneys on the sea floor that belch black fumes of superheated water.

C

calcite
One of the commonest minerals and the main constituent of limestone. Its chemical name is calcium carbonate.

carbon dioxide
A colorless, odourless gas, containing the elements carbon and oxygen, that is a part of the air people breathe. It is produced when fuel containing carbon is burned in the air. Its is also produced by breathing.

cement
A man-made powder made by strongly heating a mixture of crushed limestone and clay. When mixed with water and sand, it sets to a hard solid that is used for building work.

cirque
A deep hollow in mountains carved out by the head of a glacier.

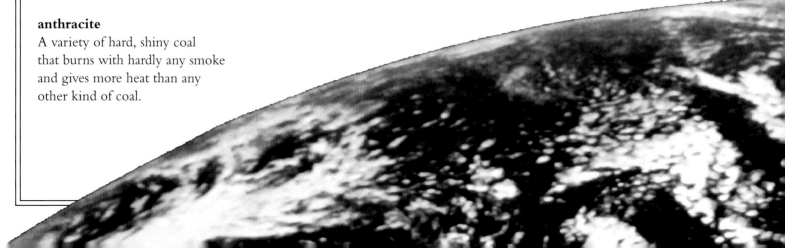

collision zone
A zone on the Earth's surface where the edge of one tectonic plate is being forced into, and under, another. It is marked by lines of volcanoes and strongly distorted rock layers.

compass
An instrument, containing a magnetized strip of metal, used for finding the direction of Earth's magnetic north.

continental crust
The thick part of the Earth's crust under continents.

continental drift
The generally accepted theory that continents move slowly around the world.

continental shelf
The zone of shallow water in the oceans around the edge continents.

convection
The rising of hot air or fluid because it is lighter than its surroundings.

core
The dense, hot metallic center of the Earth.

crevasse
A deep crack, typically in the surface of a glacier.

crust
The solid outer shell of the Earth, varying from 3–50 miles thick.

crustal plate
Also known as a tectonic plate. One of 20 or so giant areas of rock that make up the Earth's surface.

crystal lattice
The orderly arrangement of atoms inside a crystal that gives the crystal its shape and determines the angles between its different faces.

cycle of erosion
The idea that landscapes are worn down through youth, maturity and old age again and again.

D

denudation
The gradual wearing away of the landscape.

E

ecosystem
A community of living things interacting with each other and their surroundings.

element
A chemical substance, such as gold, carbon and sulphur, that cannot be further broken down into other elements.

era
A major subdivision of geological time.

erg
A large sea of sand in the desert.

erosion
The gradual wearing away of the land by weathering and agents of erosion such as rivers, glaciers, wind and waves.

F

fault
A fracture in rock where one block of rock slides past another.

fetch
The distance winds blow over the ocean to create waves.

firn
Snow compacted into ice by melting and refreezing.

fossil
The remains, found preserved in rock, of a creature that lived in the past.

G

geophysics
A sister science to geology concerned with the physical properties of rocks, such as magnetism, density and radioactivity.

glacial drift
All the material deposited by a glacier or ice sheet.

glaciation
The molding of the landscape by glaciers and ice sheets.

glacier
A river of solid ice.

gneiss
A common type of metamorphic rock produced deep in the Earth's crust. It is formed when other rocks are subjected to strong pressures and high temperatures.

goniometer
An instrument used to accurately measure the angle between two of the faces on a crystal.

greenhouse effect
The way certain gases in the atmosphere trap the sun's heat like the panes of glass in a greenhouse.

groundwater
Water that percolates downward from the surface of the Earth into spaces in the rocks below.

guyot
A flat-topped mountain under the sea, typically a volcano that has been eroded at the summit by waves, and which has then been submerged.

H
hanging valley
A side valley cut off and left hanging by a glacier.

hot spot
A place where hot pockets of molten rock in the Earth's mantle burn through the Earth's crust to create volcanoes.

hydrocarbon
A chemical compound containing the elements hydrogen and carbon.

I
ice age
A long cold period when huge areas of the Earth were covered by ice sheets.

igneous rock
A rock that forms when magma (hot, molten rock from the Earth's interior) cools and becomes solid.

impurity
A substance in a mineral or a crystal that is additional to the normal components, and which may cause variations in color. This can happen on the Earth's surface or underground.

iron oxide
A compound (mixture of elements) found all over the Earth that contains both the elements iron and oxygen in various proportions.

J
joint
A fracture in rock between two blocks.

K
karst
A type of scenery found in limestone regions where the rainfall is high. It features steep-sided pinnacles separated by deep chasms.

L
lava
Molten rock (magma) that flows from volcanoes or cracks in the Earth's crust.

limestone
A rock, usually sedimentary, formed almost entirely of the mineral calcite.

lithosphere
The rigid outer shell of the Earth, including the crust and the rigid upper part of the mantle.

M
magma
Hot molten rock in the Earth's interior, which is known as lava when it emerges on to the surface in eruptions.

magnetism
An invisible force found in some elements but especially in iron, which causes other pieces of iron to be either pushed apart or drawn together.

metamorphic rock
A kind of rock, such as gneiss, created when the structure of other rocks is altered by intense heat or pressure from inside the Earth.

meteor
A piece of rocky material from space that burns as it falls through the Earth's atmosphere, producing a streak of bright light.

mica
A common crystalline mineral, found in igneous rocks, which splits into thin, flexible, transparent sheets.

micrograph
A photograph taken through the lens of a microscope, using a special camera designed for this purpose.

mid-ocean ridge
A long jagged ridge on the ocean floor along the gap between two tectonic plates which are moving apart.

mineral
A naturally occurring substance, found in rocks.

moraine
Sand and gravel deposited in piles by a glacier or ice sheet.

N
nappe
A complex mass of folded-over rock strata in mountain regions.

O
ore
A mineral from which a useful material, especially metal, is extracted.

orogeny
A major period which fold mountains were formed by the movement of the Earth's tectonic plates.

ozone
A form of oxygen gas. It is poisonous, but a layer of it high in the stratosphere protects us from the Sun's harmful ultraviolet radiation. The ozone hole is where the ozone in the stratosphere is very sparse.

P

palaeontology
The study of fossils.

Pangea
A megacontinent of about 220 million years ago that cracked apart to form all today's continents.

permafrost
Permanently frozen ground, in which temperatures are below 32°F for more than two years.

pigment
A finely powdered substance used to give color to a material without dissolving in it, unlike a dye.

plaster
Finely powdered mineral, commonly gypsum (or calcium sulphate), which is mixed with water to make a paste for smoothing walls inside buildings.

plate
See tectonic plate.

porcelain
A fine, semi-transparent pottery made by firing a mixture of kaolin (a type of clay) and various other ingredients at high temperature.

Q

quarry
A large, man-made, open hole in the ground from which minerals are taken.

R

recycle
To convert something old into something new. In nature, old rocks are continuously being changed into new rocks by movement in the Earth.

rift valley
A valley formed when a strip of land drops between two faults.

S

sedimentary rock
A rock made up of mineral particles that have been carried by wind or running water to accumulate in layers elsewhere, most commonly on the beds of lakes or in the seas and oceans.

seismograph
A sensitive instrument used to detect earthquake waves, from the slightest tremor to powerful shocks.

soil
Material produced from rock, at the surface of the Earth, by the action of the weather, plants and animals.

specific gravity
A number used by scientists to indicate how heavy (or dense) a material is, no matter how large or small is the sample.

strata
Layers of sedimentary rock.

stratosphere
The layer of atmosphere above the troposphere where temperatures get warmer higher up.

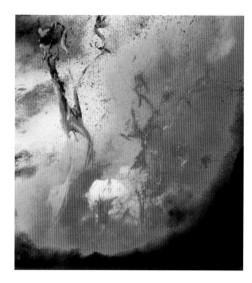

subduction
The bending down of one tectonic plate beneath another as they collide.

syncline
A dish-like downfold in layers of rock.

synthetic crystal
A man-made crystal identical to one found in nature.

T

tectonic plate
The 20 or so giant slabs of rock that make up the Earth's surface.

thermosphere
The layer of the atmosphere above the mesosphere, beginning 50 miles up.

till
Mixture of rock debris left by an ice sheet over a wide area as it retreats.

tor
A clump of big blocks of bare rock on top of a smooth hilltop.

transverse dune
A dune at right angles to the wind.

troposphere
The densest, bottom layer of the atmosphere, up to 7½ miles.

W

weathering
The breakdown of rock when exposed to the weather.

INDEX

A
acicular 15
acid test 32
agate 60
aging rocks 31
aggregate 48
algae 76–7, 118
alloys 55
aluminum 12, 55
amber 44, 45, 61
amethyst 15
amino acids 76
ammonites 9, 45, 78
amphibians 76, 79
andesite 57
anthracite 50
antiforms 24
aquamarine 60
aquariums 116–17
aragonite 15
Armstrong, Neil 67
asteroids 59
atmosphere 66, 70–1, 74–5, 88, 108, 118–19
atolls 103
atoms 14, 16
attrition 42
auroras 71

B
barite 15
barkhans 40
basalt 12, 18–19, 59
bauxite 55
beaches 8, 23, 40, 41, 98
beryl 33, 60
bitumen 49
black smokers 76
blue john 61
brick 49, 56
bronze 55
building materials 48–9, 56–7

C
calcite 14, 28, 32, 38–9, 61
cameo 60
Cararra marble 56, 57
carbon 13, 55
carbon dioxide 77, 108, 119
carbonates 30
caves 36, 38–9
cement 48
chalcopyrite 55
chalk 8, 22, 38
chondrite 58
chromium 61
citrine 15
clay 8, 22, 48–9, 52
cliffs 8, 22, 23, 24, 38, 98–9
climate 40, 110–13
climate change 108–9, 119
clouds 75, 106
coal 49, 50–1, 79, 119
   coastlines 98–9
   color 32, 60, 61, 62
compasses 72–3
concrete 48, 56, 63
conduction 54
conglomerate 22, 26–7, 35
contact metamorphism 28–9
continents 66, 70, 82, 84, 100, 103, 105, 108
coral 35
coral reefs 103, 109, 114
corundum 61
craters 58
crevasses 94
crust 70, 82, 84
crustal plates 34
crystal lattice 14, 15
crystals 8, 9, 12–17, 18, 20, 29
   twinned 15
currents, ocean 102, 104–5

D
Darwin, Charles 80, 117
decomposers 43
deltas 23, 41
dendritic 15
dendrochronology 110–11
density 31, 32–3
denudation 90
desert rose 15
deserts 40, 42, 89, 96–7, 106
diamond 8, 13, 14, 60–1
dinosaurs 77–9
dolerite 18
dunes 40, 96

E
Earth 8
   core 36
   crust 9, 18, 23, 26, 28–9, 30, 34
earthquakes 36, 82–3
ecosystems 114–17
elements 12–13
emerald 13, 60
equator 67, 69, 84, 102, 105–7, 112–13
erosion 22, 38, 40, 88–91, 97–9
evolution 78–80
exosphere 75
extinction 119
extrusive rocks 18

F
faults 24, 26, 85
feldspar 12, 19
fjords 95
flint 38, 48
floodplains 92
floods 96, 115, 119
fluorite 61
fold mountains 86–7
folds 24, 26
foliated 28
fool's gold 55
forests 112–13, 115–16, 118
fossils 9, 23, 28, 30, 31, 34, 44–7, 50, 76, 78–9, 109, 110
freestone 56
frost 88, 90

G
garnet 29
gases 8, 30, 32, 74–5
genes 81
gemstones 8, 13, 14, 60–1
geodes 9
geology 30–1
geophysics 36–7
glaciers 90, 94–5, 108
glass 14, 48–9
gneiss 9, 28, 29, 34
global warming 108, 119
gold 12, 13, 54–5, 62
goniometer 17
graded beds 43
granite 9, 12, 18–19, 49, 56, 57
grasslands 112–13
gravity 36, 70, 75, 100–2
greenhouse gases 108, 119

grikes 39
groynes 98
gypsum 15, 48
gyres 102, 104

H
habit 15
haematite 22
halite 14, 23
hardness, testing for 11, 32
heliodor 60
hematite 15, 55
hornblende 12
hot springs 71
humans 77, 118–19
humus 43
hydrocarbons 52–3
hydrogen 13

I
ice 16, 66, 75, 88–90, 94–5, 108–9, 112
Ice Ages 77, 94–5, 98, 108, 110
identification 32–3
igneous rocks 9, 12, 14, 18–21, 56
intrusive rocks 18
iron 8, 22, 49, 54–5, 58, 70–2
islands 66, 83, 103

J
jade 61
jadeite 61
jet 51
Jupiter 59

K
kaolin 49
karst 38–9
kimberlite 13
kopjes 89
kyanite 8

L
lakes 92, 116, 118
lamellar 15, 29
lapis lazuli 61
lava 9, 18–19, 59, 86
lazurite 61
life 66, 74, 76–81, 114–16, 119
lignite 50
limestone 22–3, 28, 35, 36, 38–9, 45, 48–9, 56, 57, 89–90
limestone pavement 39
liquids 8, 12, 16, 52–3
lithosphere 23, 82
luster 32

M
Magellan, Ferdinand 66
magma 9, 14, 18–19, 20, 28–9, 70, 85
magnetism 34, 36, 58, 71, 72–3
magnetosphere 71, 73
mammoths 79
mantle 70, 82
maps, geological 30
marble 28, 49, 56, 57
Mars 5, 58, 59
meanders 92, 93
Mercury 66
metals 13, 16, 54–5
metemorphic rocks 9, 28–9, 56
meteorites 58–9, 70, 74, 108
meteoroids 58–9
mica 12, 15, 28
mid-ocean ridges 83, 102–3
Milankovitch cycles 109
minerals 8, 12–13
minerology 30
mining 36, 50–1
Mohs' scale 11
monoliths 22
monsoon 107
Moon 58–9, 67, 69–70, 74, 88, 100–1
moraines 95
morganite 60
mountains 8, 24, 34, 82, 84–8, 90, 92, 94–5, 113, 116
mudstone 23, 28, 45
muscovite mica 15

N
nephrite 61
nitrogen 74–5
North Pole 72, 106
Northern Hemisphere 68, 102

O
obsidian 18
oceans and seas 66, 74, 98–105
oil 48, 52–3
olivine 19, 28, 58
opal 8
orbit 109
ores 13, 54–5
oxygen 12, 13, 71, 74–7
ozone layer 118–19

P
paleontology 28, 44
Pangea 82, 84
peat 40, 50
pebbles 23, 26–7, 35
permafrost 95
petrology 30, 31
pigments, paint 60, 61, 62
planetismals 70
plants 71, 76–80, 112–14, 116, 118
plutons 18–19
polarized light 13, 31
poles 66, 71–2, 102, 105–6, 108, 112–13
pollution 108, 118–19
Portland stone 57
potholes 39
puddingstone 22
pyrite 8, 14, 15, 33, 55, 61
pyroxene 12, 19, 58

Q
quarries 8, 9, 29
quartz 12, 14–15, 22, 40, 41, 48

R
radioactivity 31, 36
rain 88–90, 92, 97, 106–7, 118
rainforests 113, 115–16, 118
reniform 15
rhyolite 18, 57
rift valleys 85
rivers 8, 23, 24, 41, 52, 88–9, 92–3, 97, 118
rocks 8, 12
  coastlines 98–9
  erosion 88–90
  formation of Earth 94, 97
  fossils 76–7
  tectonic plates 82–3
ruby 13, 14, 60, 61
rust 54

S
salt 14, 22, 23
sand 8, 40, 41, 48–9, 96–8
sandstone 9, 22, 31, 35, 56, 57
schist 29
scree slopes 88

seas and oceans 66, 74, 98–105
seasons 68–9
sedimentary basins 35
sedimentary rocks 9, 22–7, 35, 38, 44, 50, 56, 86
sediments 93, 110
seismic reflection survey 37
shale 28, 35
shells 25, 35
shrimps 80
silicates 12
sky 74
slate 28, 29, 56
smelting 54–5
snow 92, 94, 106, 108–9, 112
soil 40–3
solar system 66, 70
solar wind 73
solids 16
South Pole 72, 106
Southern Hemisphere 68, 102
species 78, 80, 114, 119
specific gravity 32–3
stalactites and stalacmites 38–9
stars 69–70
storms 99, 104, 106–8
strata 23, 24–5, 26
stratosphere 74–5, 119
streak test 32
stromatolites 77
sugar 16, 20
sulphur 8, 12, 13, 30, 61
Sun 66–71, 74, 101, 106, 108–9, 119
sunspots 108
synforms 24
synthetic crystals 13

T
taiga 112
tectonic plates 82–3, 85–6
tides 98–101
tors 89
transparency 32
travertine 38–9
trees 77–8, 112–16, 118
trenches, ocean 82
troposphere 74–5
tundra 95, 112

V
valleys 8, 24, 25, 85, 88–9, 92, 94, 95
vegetation zones 112–13
veins 9, 14, 28
volcanoes 18, 30, 59, 70, 71, 76, 82–4, 86, 88, 103, 108

W
water 8, 52, 66
  aquariums 116–17
  atmosphere 74
  deserts 96
  erosion 88–91, 97–9
  oceans and seas 66, 74, 98–105
  rain 88–90, 92, 97, 106, 107, 118
  rivers 88–9, 92–3, 97, 118
  sediments 95
  tides 98–101
  tundra 95
  vegetation zones 112
waterfalls 92
waves 98–9, 103–4
weather 88, 90, 106–8, 110–11
weathering 40–1
wetlands 115
winds 88–90, 96–7, 102–7

ACKNOWLEDGMENTS

The publishers would like to thank the following children for modeling in this book: Emma Beardmore, Mitchell Collins, Joshua Cooper, Ashley Cronin, Joe Davis, Louise Gannon, Hamal Gohil, Sarah Kenna, Catherine McAlpine, Griffiths Nipah, Goke Omolena, Ben Patrick, Daniel Payne, Anastasia Pryer, Charlie Rawlings, Kristy Saxena, Georgina Thomas, Victoria Wallace and all the children and staff of Hampden Gurney School. Also: Mr and Mrs G R Evans, Ian Kirkpatrick. A special thank you to Gregory, Bottley and Lloyd for their efficient help in supplying rock and mineral samples.

PICTURE CREDITS (b= bottom, t= top, c= centre, l= left, r= right)
The Art Archive: 54bl, 60br; BBC: 77tl, 80c, 86tr, 88br, 93c & bl, (also 100tr), 97t & c, 99t & b, 107t & c, 112tl + 1 reused in glossary; BBC Natural History Unit /Pete Oxford:62cr; Bridgeman Art Library: 56tr, 60cr, 61br; British Atlantic Survey: 26t; British Antarctic Survey: 31cr, 37bl & c; British Geological Survey: 25cr, 31tl, 41bl, 50br; Thomas Chatham: 13br; Bruce Coleman Collection: 83br, 92br, 109c, 112c, 119tr; /S. Bond: 19tr, /J. Cancalosi: 45bl, /J. Cowan: 39bl, /D. Croucher: 19bl, /J. Foott: 19tl, /C. Fredriksson: 40b, /G. Harris: 39br, /J. Shaw: 60tl; Corbis: 75bl, 78tl, 79c, 109t, 110tl, 112c +1 reused in glossary; Sylvia Cordaly: 105c; De Beers: 8tr, 11cr, 60bl; Digital: 54br; Ecoscene: 84c; Genesis: 67tl, 74br; GeoScience Features Picture Library: 12cl, 14bl, 18tr, 29cl, 33cl & bl; Robert Harding: 37bl, 38bl; /V. Englebert: 30bl & br, /N. Francis: 54bl, /P. Hadley: 48br, /L. Murray: 31tr, /Tony Waltham: 51cl; Frank Lane Picture Agency: 48cl, 50bl; /L. Batten: 35tl, 39tl, /C. Carvalho: 24br, /S. McCutcheon: 24bl, 27bl, /C. Mullen: 56bl /M. Newman: 22tr, /M. Niman: 18tl, /M. J. Thomas: 19br, /R. Tidman: 23br, /T. Wharton: 41tr , /W. Wisniewski: 40tl, 50tr; Microscopix: 13bl, 31tr, 41cr & br; /A. Syred: 31bl; Milepost 9½: 51bl; NASA: 23tr, 45br, 58tl & cl, 59tl, cl & cr; Natural History Museum: 9cr, 25tr, 50l & c; Natural History Photographic Agency (NHPA): 46c, 77t, 83tl, 85t, 88tr, 89c, 99c, 101c, 108tl, 111c, 113c, 114t, 116tr, 117c; /G. Bernard: 29cr, /D. Woodfall: 50bl; Oxford Scientific Films: 81t, 112b; /B. Herrod: 53cr, /R. Packwood: 57br, /D. Simonson: 57cl, /K. Smith Laboratory & Scripps Institute of Oceanography: 35br, /M. Slater: 57tl, /S. Stammers: 44c, 47cr, /H. Taylor: 35cr, /R. Toms: 52tr, /G. Wren: 40cr; Papilio Photographic: 25b, 85bl, 115c, 118c, 69b; Planet Earth Pictures: 66c, 66tl, 67cr, 70tl, 71br, 75c, 78c, 78bl, 79c, 82br, 84tl & br, 89c, 92tr, 93t, 94b & c, 95tr & c & br, 96tl & br, 97bl, 98tr & br, 102tl, 102tr & br, 106tr & c, 113c, 114b, 118c & b, 119tl; Powerstock/ Zefa: 48bl; Science Photo Library: 14tr & cl, 76t, 108cl & cr, 119c; /M. Bond: 29b, /T. Craddock: 52br, /Crown Copyright, Health & Safety Laboratory: 55br, /E. Degginger: 13cl, /M. Dohrn: 23tl, /S. Fraser: 34b, /F. Gohier: 59br, /J. Heseltine: 34tl, /M. McKinnon: 53bl, /A. & H. Michler: 58c & l, /NASA: 58br, /P. Parriainen: 58c, /A. Sylvester: 19cl, /S. Terry: 57br; Skyscan: 57tr, 75br; Tony Stone: 51bl & br, 56br, 71t, 72tl, 75t, 77c, 79bl, 90tr, 95tl, 103tl, 107b; Trip /J. Arnold: 38br, /Phototake: 13t; University College London /A.R. Lord: 44l; University of Glasgow /Dr Gribble: 17c.

Every effort has been made to trace the copyright holders of all images that appear in this book. Anness Publishing Ltd apologises for any unintentional omissions and, if notified, would be happy to add an acknowledgment in future editions.